THE PERFORMANCE CORTEX

"To use a voguish sports catchphrase, *The Performance Cortex* is 'next level.' We've heard a lot about 'mental toughness' and 'hard-wiring for success,' but now Zach Schonbrun reveals the latest science on how elite athletic feats are actually accomplished. Fans will understand the genius behind all sports more clearly after reading this book. And they can, with pleasure. Schonbrun has mastered the art of writing gracefully about dense—and potentially groundbreaking—material."

—L. Jon Wertheim, executive editor of *Sports Illustrated* and coauthor of *This Is Your Brain on Sports* and *Scorecasting*

"Zach Schonbrun's *The Performance Cortex* is full of insight into the next wave of athletic training, the relationship between the mind and the body, and the cutting-edge neuroscience that seeks to explore and exploit this interaction to create better athletes. This accessible account will leave every reader wishing they had known all this before."

—Glenn Stout, author and series coeditor of *The Best American Sports Writing*

"The brain is the last untapped resource for athletes, the final frontier for sports analytics. Zach Schonbrun's riveting look inside of how players' minds truly work, and how that knowledge is being used to reimagine the games we play, fires with the efficiency and efficacy of a synapse."

—Jeff Passan, national baseball columnist for Yahoo! Sports and author of the *New York Times* bestseller *The Arm*

The
PERFORMANCE
CORTEX

HOW NEUROSCIENCE IS
REDEFINING ATHLETIC GENIUS

Zach Schonbrun

DUTTON

DUTTON
An imprint of Penguin Random House LLC
penguinrandomhouse.com

Previously published as a Dutton hardcover edition
First trade paperback edition: April 2019

THE LIBRARY OF CONGRESS HAS CATALOGUED THE HARDCOVER EDITION
AS FOLLOWS:
Names: Schonbrun, Zach, author.
Title: The performance cortex : how neuroscience is redefining
athletic genius / Zach Schonbrun.
Description: New York, New York : Dutton, [2018] |
Includes bibliographical references and index.
Identifiers: LCCN 2017048817 (print) | LCCN 2017050826 (ebook) |
ISBN 9781101986349 (ebook) | ISBN 9781101986332 (hc)
Subjects: LCSH: Sports—Physiological aspects. | Athletes—Physiology.
Neurophysiology. | Cerebral cortex—Physiology. | BISAC: SPORTS &
RECREATION / Sports Psychology. | SCIENCE / Life Sciences / Neuroscience. |
SPORTS & RECREATION / Training.
Classification: LCC RC1236.N47 (ebook) | LCC RC1236.N47 S36 2018 (print) |
DDC 612/.044—dc23
LC record available at https://lccn.loc.gov/2017048817

Dutton trade paperback ISBN: 9781101986356

1   3   5   7   9   10   8   6   4   2

Book design and illustrations by Daniel Lagin
Set in Warnock Pro with Conduit ITC Pro

147429898

To my parents.
And to Missy, my reason for movement.

CONTENTS

CONTENTS

THE PERFORMANCE CORTEX

INTRODUCTION

L ast spring, I traveled to Dublin to attend the Society for the
Neural Control of Movement's annual conference, where I
heard, on day one, a presenter elegize the recent passing of a
dear colleague. "He was never happier than when he was descend-
ing electrodes into the spinal cord looking for a neuron," he said.
"When he found one, he treated it like the first neuron he found." A
lesson therein for us all. I myself was not sure what exactly I had
descended into. One attendee, Elżbieta Jankowska, began her career stimulating the lumbosacral region of decerebrate cats more
than half a century ago. Another claimed to be the "academic great-
great-grandson" of Claude Bernard. One of the most decorated active researchers in the world, Tom Jessell, was there to discuss his
work with mouse genetic tools. And me? I was there because of
my wife.

She discovered the small blurb in my Columbia University

alumni magazine about the two neuroscientists trying to work in Major League Baseball. I knew of sports psychology, mindfulness training, even brain gaming as a growing fad among professional franchises. But *neuroscience* seemed to represent a different level of sobriety. What were they looking for? What had they found?

I met Jason Sherwin at a dingy Jamaican buffet with a bright crimson awning in East Flatbush, across from the SUNY Downstate hospital where he was working. We still joke about the "mystery meat" served alongside collard greens and gummy plantains. We spoke for two hours as he related his life, his résumé, how he met Jordan Muraskin, how he envisioned their company as helping to usher in "Moneyball 2.0": biometric analytics, a priori probabilities, brain data. I wrote the story for the website SB Nation Longform, a now-defunct outlet for sportswriting's deep cuts. But as I wrote it, I knew it was a sports article by its place of residence only. On the surface, the efforts by Jason and Jordan were intended to help professional baseball teams scout and improve hitters. But, to zoom out a bit, their endeavor seemed to be more like tracing the essential correlates of a skill. This skill could be anything that requires a rapid decision: passing to an open wide receiver, whistling a foul call, responding to gunfire after a report of breaking and entering. Those are outcomes, like the speed of a car as it zips down the highway. Jason and Jordan encouraged me to reconsider what is going on beneath the hood. Hitting a baseball, to take one of the more straightforward outcomes, has been deemed "the most difficult thing to do in sport." Some might quibble about this, but those who do normally have not tried it. The most proficient hitters are hardly at all cut from the same cloth. The two front-runners for the Most Valuable Player of the American League in 2017, in fact, were a

Venezuelan infielder standing five feet, six inches tall and weighing 165 pounds (José Altuve) and a Californian outfielder standing six feet, seven inches and weighing 282 pounds (Aaron Judge). We already know what distinguishes them; we can see it. So what relates them? What actually is responsible for their skill? Jason and Jordan wondered, and so did I, once I really started to think about it. It would seem to have nothing to do with their biceps muscles or fast-twitch fibers or even their vision, which for most baseball players is largely the same. It would seem to have much more to do with the neural signals that impel our every movement.

How do we move? A few people have looked into this. The Egyptians actually wrote of head injuries and movement disorders. Erasistratus and Herophilus explored the cerebella of fast-moving animals like deer and rabbits. Galen of Pergamon learned about the brain from tending to the wounds of the gladiators. The origin of movement had bewitched some of history's shrewdest minds: Alcmaeon, Plato, Aristotle, Posidonius, Al-Razi, Descartes, Newton, Franklin. When the brain's primary seat of voluntary action, the motor cortex, was finally discovered, by a pair of wayward Germans in 1870, the operation had been conducted on a dog sprawled across a dresser at the home of one of the men. The eureka moment howled from a living room in Berlin.

Since then, most of the motor research has been conducted more quietly. For some reason, the field does not seem to attract the attention given to clinical tales or the various meditations on our cognition, such as the neuroscience of self; the neuroscience of language; the neuroscience of dogs; the neuroscience of consciousness; the neuroscience of being a good parent. An Amazon book search on the phrase "neuroscience of movement" turned up a fuchsia textbook by a professor of physical therapy published in 1997 and scant other options. Part of the problem might be that the question of movement sounds old and elemental, the stuff of anatomy classes and collapsible polyvinyl skeletons. The other problem is that progress has been a bit slow. When I visited one neuroscientist, he was in the middle of crafting a rebuttal to a controversial interpretation of motor learning. The offending interpretation was made in 1951. A popular experimental paradigm in most motor research labs is called a "force-field adaptation task," which was first introduced in 1994. It replaced the reciprocal protocol task, born in 1954. Other techniques came and went. Theories appeared and disappeared like pimples. As I walked outside the conference hall at the Clayton Hotel during the NCM meeting, I scanned the bewildering titles on hundreds of posters, each being presented with hand waves and speeches in fast-forward by nervous postdocs. "All of this is going to change," said Jordan Taylor, a professor at Princeton, waving his own hand toward the rows. Maybe he was right. I hoped to capture it while it lasted.

I ordered a textbook called *Principles of Neural Science*. It arrived weighing almost 10 pounds, checking in at 1,760 pages. Skimming through it, I learned that it takes a tenth of a second to process everything we see; that a newborn infant is able to reflexively churn its legs, despite the fact that its spinal cord cannot yet transmit messages down from the brain; that visual inputs get siphoned into two streams, a "where" and a "what"; that the "where" stream is also sometimes called the "how" stream, and the "what" stream imbues our inputs with meaning; that the sense of touch on our fingertips is twice as bad at age 70 as it was at age 20; that the Behaviorists came before the Cognitivists. This was just Chapter 38. By now I had a book on my hands. The first person I called was neuroscientist John Krakauer, a silver-tongued polymath I had come across, somewhere, haranguing a hapless science reporter about Michael Jordan and behind-the-back passes. "I think you just need to decide, what kind of piece do you want to write?" he told me. "Do you want to write about the 'motor system hunters' and what they're beginning to learn, or do you want people to speculate about what makes these top athletes so good?"

I hung up the phone in a daze. All my life I had admired athletes, fantasized about being one myself, and in my professional life I had been fortunate to get up close to many of the greatest. But one forty-minute phone call had irrevocably pierced my shroud of ignorance. All my life I had focused on the body. I realized now that my attention had been amiss. "It's like saying people who can speak French very well have a very dexterous tongue," Krakauer said. "It would be the wrong place to assign the credit."

As I proceeded to spend more time with Krakauer, and then with Adrian Haith and Daniel Wolpert and Emily Cross, Jörn Diedrichsen, Andrew Pruszynski, Doug Crawford, Dagmar Sternad, Bob Kirsch, Daniel Laby, and many others, a clearer picture of the story I wanted to tell began to form. Readers of this book might be surprised not to find much discussion of two popular and seemingly relevant topics: genetics and concussions. There are more than enough books devoted to each of those subjects, and more undoubtedly on their way. Instead I endeavored to stay faithful to an area that had been, in my opinion, woefully underserved: the motor system. I would focus on the men and women, contemporary and across history, who have devoted their lives to understanding how the motor system produces the performances we watch and adore.

The narrative would remain anchored by the baseball diamond, to that purest of athletic exchanges, when a batter stands at the plate awaiting a pitch. All along, I, and millions of others, had cared solely about the infinite possible outcomes that could result from that confrontation, rather than the infinitesimal interactions in the four-tenths of a second in between. It was time to give those milliseconds their due.

1.

DECERVO

"HOW CAN YOU THINK AND HIT AT THE SAME TIME?"

There was no indication that anything unusual was taking place on an early Saturday morning in August at the Hilton Garden Inn, of Avondale, Arizona, other than the piece of loose-leaf paper taped to the wall by the elevator bank. On it was scribbled in black Sharpie: DECERVO TESTING ROOM 307. The room number was underlined. The tone was "no trespassing." Still, the housekeeper knocked on the door of Room 307 at 8:15 with an armful of fresh towels. No one answered, so she used her key to enter. When she did, she did a double take. The furniture in the dumbbell-shaped suite had been rearranged completely. The beds were still made and the blinds were drawn. Two scrawny, acne-pocked Latino teenagers in T-shirts and sandals were seated at matching desks on opposite sides of the room staring unblinking at laptop screens. Each wore a sort of thin metallic hairnet, with wires snaking down the back of their necks. A pile of plastic syringes and two padded briefcases lay scattered on the floor. The only sound came from soft, intermittent taps on the laptop keyboards. Neither of the men looked up to see the housekeeper quickly drop the towels off and go.

In the everlasting war for even the slightest competitive advantage in Major League Baseball, the battlefields have come to look a lot different than the playing fields. They have left the playing fields behind. This new terrain was once thought to be impregnable. Now, suddenly, held captive on Saturday mornings in suburban hotel rooms, it was spilling its secrets. When other teams learn of this, they will undoubtedly try to do the same. "Moneyball" was that way; once the data-driven revolution started, it became difficult to contain, until every team started using advanced analytics to discover new players or rediscover old ones. Then the battle had to be moved someplace else. Those teams that were late to that data revolution now had a chance to get ahead in this one. This data revolution required a new type of radar gun, one that could measure in milliseconds.

At 8:25, there was another knock at the door of Room 307. A third baby-faced teenager appeared: Manny, a shortstop, wearing a gray T-shirt and sandals, his eyes puffy and reddened. The boys, they were really just boys, had played in a doubleheader the day before, in the searing Sonoran heat, as the playoffs neared. This being a rookie-league team, below Single-A, even below Low-Single-A, every player had recently been drafted or acquired from overseas. It was their first taste of professional American baseball. They remained years away from a whiff of a chance at the Majors; most will never even get that. But as the newest and youngest parcels of a Major League Baseball empire, they are handled delicately. They reside in the hotel, a short drive from a hulking, concrete-and-glass Spring Training complex, where they relax and train in uniforms that bear the familiar colorway of their big-league parent club. They are currently chaperoned by Frank, the organization's director of sports science, who popped in and out of Room 307 with a list of the telephone numbers to each of the players' rooms, in case any of them tried to sleep in. A stocky man with soft blond hair, reddish cheeks and bright eyes, he is friendly, but with a no-bullshit mien, like a waiter at the end of his shift. Frank did not seem to care that 8:00 A.M. for an 18-year-old on the Saturday morning after a late doubleheader is a considerable, if not downright malicious, request. But there was a lot to get done. Jason Sherwin and Jordan Muraskin were only in town for two days. Their sort of expertise is not easily replaceable. The club paid $2,000 to fly them out there. As the ballplayers tapped on their keyboards, and Manny waited on the couch, Jason and Jordan shushed about, adjusting the hairnets. They chatted idly with Frank about the upcoming fantasy football season, but there were giveaways that they were not members of a typical athletic entourage. Noticing the colorful symbol on the front of Manny's T-shirt, Jordan asked him, "Is that a Google shirt?" "No," he replied sheepishly. "World Baseball Classic."

A spot opened up at one of the desks after the first player finished. Manny took a seat and waited as Jason prepared the laptop and Jordan readied the headgear. He used an alcohol swab to dab behind the player's ears and fitted the strange translucent swim cap—an EEG headset—over Manny's short hair. Then he grabbed a syringe and squirted a pale creamy substance into the seams around the nine spots where the sensors were expected to maintain the closest contact with the skull. The cream, the consistency of toothpaste, is a conductive salve for the electrodes. "You remember this?" he asked. Manny nodded. He quickly typed his username and password into the system and the screen went dark, with only a small rectangular box appearing in the center. A moment later Jason signaled the program was ready.

"It'll take about 40 minutes," Jordan said. "Do you want any practice?"

"No," Manny said. "I'm good."

The simulation began. "And we're off," Frank said.

First came the orthopedists. They came to baseball in the late 1950s and early 1960s, transforming how pitchers were assessed and treated. The psychologists followed. Then the optometrists, strength coaches, massage therapists and nutritionists. The economists and sabermetricians. The Zen masters and sleep doctors and yoga instructors. And finally, at last, there came the neuroscientists, fresh from school, brandishing doctorates and peer-reviewed papers and exactly nothing of any value that mattered to a baseball executive other than their answer to the question, "Can you make my team better?" To which Jason and Jordan, cofounders of a startup called deCervo, would answer unequivocally, "Maybe." They were not sure. They were scientists, and they had no background in business. They seemed to have no business in baseball. That a sport moored to tradition—where managers still wear uniforms in the dugout and make calls to the bullpen using a landline phone; the last major league to adopt on-field instant replay—had any interest in doing business with them was also unclear. But they wanted to help. There is a saying about baseball that, even after 170 years, you can still see something new in any game. Outside of cheating, though, there was not much new for improving the act of hitting. Hitters can so often seem besieged by so many factors—mounting velocities, defensive shifts, the un-yielding constraints of our visuomotor system—that reaching base safely even on occasion is widely considered a paragon of skill. Ted Williams once called hitting a baseball "the most difficult thing to do in sport." Some say that the hands need to be "coordinated" well with the eyes, which can be deceiving if you are one to believe that hitting, like a lot of athletic endeavors, is mainly a rote exercise based on muscle memory, a term coaches use often. Coaches also say things like "watch the ball hit the bat" and "slow the game down." In 1921, psychologists at Columbia University designed a battery of sensory-motor tests for Babe Ruth, under the guise of empiricism, to explain his prodigious hitting ability. After the tests, the research-ers declared (misleadingly) that Ruth's vision, reaction time and coordination were far and away better than others'. It is unknown if any other ballplayers were assessed. The study was headline news at the time. A hypothesis for what it takes to be a great hitter emerged: Be Babe Ruth.

But almost a full century later, two other researchers from Co-lumbia University began to try a different approach, this time with techniques adopted from a place called the Laboratory for Intelligent Imaging and Neural Computing. Their assessments probed deeper than the cursory physical examinations performed on Ruth. They went hunting for data. *Brain data*. With the EEG and a carefully tai-lored video simulation, Jason and Jordan believed they had landed upon a novel collection method for the type of information previ-ously left to guesswork. They could transport their services any-where, taking their exploratory findings out of the confines of the laboratory and bringing them into the dugouts of Major League

Baseball. It was relatively quick, entirely painless, and no more invasive than using an Apple Watch to measure your heart rate.

When the two researchers were first testing the headgear on players for the baseball team at Bradley University, in the fall of 2014, they finally got the full scope of the concept they were developing and what it could mean for baseball. The coaches wanted to know the results of one player in particular, who seemed to struggle at the plate despite obvious athleticism and a picturesque swing. Jordan, then 29, slender, with a boyish face, trimmed dark hair and strong eyebrows, patiently explained the metrics he had recorded during the player's neural assessment, which they categorized with academic patois like Neural Decoding Performance, Decision Position Metrics, and Neural Discrimination Strength. On Jordan's laptop, the readout was filled with line graphs, column graphs, data tables and heat maps. He explained to the coaches that the player's "neuronal curve has shifted backwards": He was late recognizing certain pitches and therefore late in deciding whether to swing.

There was silence on the other end of the line. Finally, one of the coaches said, "We never understood why he's not the best player on the team." Now they had a clue. "It was like, 'Yes, yes, yes!'" Jordan told me later.

What Jason and Jordan were showing was a baseball version of what is known as rapid perceptual decision-making, which is obviously quite different than the kinds of decisions we mull over (What should I have for lunch today?) or expressly calculate (Which exit off the highway should I take?). But the ability to hit is often mistaken for reaction time, which is virtually the same for everybody. We make fast decisions all the time on a day-to-day basis. They can be reduced, like most things related to the brain, to the patterns of spatially and temporally distinct and interdependent neuron activations. Baseball players, the really good ones, produce or respond to these activations in ways different from other people. The result is they can recognize certain pitches the same way automobile enthusiasts can recognize the make and model of a car as it disappears out of sight, or the way bird-watchers can detect an instantaneous flash of color or flight pattern. It is similar to the way a chess master can quickly visualize and interpret movements on the board. We have always known this, more or less tacitly, from quotidian statistics such as batting average or on-base percentage, which have been used to assign value to players for decades. But these, deCervo likes to point out, are post hoc variables. They come only after the player has finished his at-bat. They don't consider how much luck is involved in inflating or deflating those statistics, such as whether the ball skipped off a fielder's glove or the wind shifted direction to rescue a fading bloop. There are advanced analytics that help factor in some of that, but they are intensive and complicated. They are not always computed in the Minor Leagues and rarely get weighed by scouts scouring high schools or overseas for future big-league talent.

But from their earliest trials with the varsity teams from Bradley, Brown and Columbia, deCervo could produce graphs that pinpointed when the batter *decided* to swing versus when he decided to take, along the time line of the pitch, down to the millisecond. A hitter stands at the ready, sees a 90-miles-per-hour slider come toward him, and makes no movement of the bat. DeCervo could still delineate the moment he made that choice to look at the pitch, rather than go for it. It registered as activity on the EEG. It registered as tiny explosions of neural action. After more testing, they had graphs that showed the spectrum of response times based on different pitches; graphs that assessed the batter's concentration level (based on eye movements and the corresponding flutter of brain activity) before the pitch is thrown; graphs that correlated to the part of the brain that is firing when decisions are made. After a year, they looked at the traditional batting statistics the players had produced and compared them to their neural metrics. They showed them to the coaches. "It was dead-on," Bradley's coach, Elvis Dominguez, told me. He wound up organizing his bench around whose readouts showed a better capacity for laying off pitches, which, he believed, contributed to a higher on-base percentage.

Jason and Jordan cast out for bigger fish. They published a few academic papers and opened a Kickstarter to raise funds. They leased space at a shared desk in Columbia's subterranean startup incubator. They attended the Sloan Sports Analytics Conference, hosted by the Massachusetts Institute of Technology, and handed out business cards. They got a few short write-ups in newspapers and baseball blogs, and their Twitter account grew to 137 followers (it is now 263). They outsourced a couple of web designers from Nepal and Brazil to improve their simulation and construct an app. A few Major League teams slowly began showing interest, always with an eye over one shoulder. One executive agreed to meet with Jason only if they could speak at a Chipotle across the street from the conference they were attending, so nobody would notice.

DeCervo drew its name from French, *de cerveau*, meaning "of the brain." "There are a lot of companies that say they're doing neural," Jordan says. "They're not doing neural." He was referring to the cognitive gaming companies, most of them modeled after Lumosity, which claimed (in some cases, deceptively) to improve mental performance through an app. But Jason and Jordan did not want to

14

claim any performance benefits from utilizing their system, or direct teams how and why they should use it. "We wanted to be the first company to measure the impact" of a decision to swing, Jason said, "and relate that mental side into performance outcomes." In essence, they were a data company—they had the means to quantify something seemingly thought to be incalculable: how and when hitters decide to swing. They thought that information could be of value to the teams who knew what to do with it. Jordan said, "I found a quote from Paul DePodesta," the former front office assistant with the "Moneyball" Oakland A's and the New York Mets, "that said 'the problem isn't with scouts or scouting. The problem is that it is based on a metric that is subjective, and not data-based.' What we're trying to do is go right into there and say, 'We're scouting purely on the stats.'"

Neural data became the differentiator. "Everybody knows the head game is one of the biggest components in sports," Jason said. "But not many people know how to measure it." Brent Walker, a mental training consultant formerly with the U.S. Soccer Federation, hopped on as an informal advisor. He was familiar with video game approaches to improve pitch recognition but could immediately sense this was different. "You can hook someone up with an EEG and then compare it to swing-and-miss data and other parameters of hitting," Walker said. "In an ideal world, you can eventually get to the point where you say, based on the way he reads pitches, he's never going to be successful at this level."

Conversely, Walker said, you can create a training protocol that strives to target precisely what the hitter is not handling, based on how his brain is responding to different pitches. Others more readily recognized deCervo's potential as a scouting service. As time passed following the early college trials, Jason and Jordan were retroactively seeing definitive demarcation points within the data collected from the players whose performances bore out those neural readings. There were also those who, as Walker said, did not seem to be cutting it. "There's clearly value in this," said Vince Gennaro, the president of the Society for American Baseball Research, who has become one of deCervo's most vocal advocates. "This is one more piece of information for a decision-maker on a draft choice, or a trade, or whether to keep a player on your roster." An executive from one of the first Major League teams to work with deCervo told me that the product "absolutely opens up a new lens to evaluate hitters." Despite the reticence of teams to let slip anything that might reduce their competitive advantage, their edge in the war, the word quickly got out about deCervo: By Spring Training of 2017, they had been approached by 28 of the 30 franchises in Major League Baseball.

One flew them to Arizona in June 2016 for a full round of neural assessments with 17 recently drafted players. Six weeks later, they returned to do a second round. In exchange for granting me access, the team asked that its identity not be revealed for proprietary reasons (and Frank, the sports-science director, is not his real name). In one weekend, in Room 307, deCervo collected 12 hours' worth of cognitive measurements of the hitters. They had withstood a few technical hiccups with the EEG as well as a few player tendencies that could have corrupted the data. More than once, Jordan had to remind a player to continue using the same hand to tap the keyboard throughout the session, as switching hands could affect the signals being recorded.

By 11:25 A.M., after their second three-hour session, they had wrapped up the last trial of the season. In the hotel parking lot, they loaded the laptops, EEG cases, syringe bags, and a power strip into the back of a black Jeep SUV. "Season One complete," Jason said, slamming the trunk shut.

Jordan looked down at directions on his phone. They needed to get back to the team facility for a wrap-up meeting, but Jason was not willing to let the moment go unrecognized. As the Jeep raced west along the highway, headlong toward another appointment on the uncertain rim of an untrodden frontier, he pounded the steering wheel with his palms. "We need to revel in this moment," Jason said. "Maybe we'll go in front of the Indians facility and do a dance like Wesley Snipes in *Major League*. Yeeaaaaowwwww!!"

In 1993, while drinking beer and smoking cigarettes at a Florida restaurant during Spring Training, Phillies outfielder John Kruk was approached by a stranger who told him his habits seemed unbecoming of a professional athlete. "I ain't an athlete, lady," Kruk replied. "I'm a baseball player."

Kruk went on to hit .316 in 1993. The line became the title to his autobiography. He had an ample waist and limited range at first base, but the Krukster could always hit. In 10 Major League seasons, he was a three-time All-Star, with a lifetime batting average of .300, right between Mickey Mantle (.298) and Willie Mays (.302). Baseball is an oppressive sport where even the best hitters make outs seven out of every 10 at-bats. But the length of the season and number of at-bats largely rules out the aspect of chance. While basketball centers can hover near the basket, and tennis players can overwhelm with a serve, baseball hitters are playing one against nine in every appearance. There is a perceived threshold for how good they can be

(.400) as well as a floor for how bad they can be before patience runs out (.200). The difference between .400 and .200 can seem like the difference between a skateboard and the USS *Nimitz*.

Then, in the spring of 1994, Michael Jordan decided he had had enough dominating basketball and wanted to try his hand at baseball. At the time, Jordan was almost universally considered the greatest athlete in the world, and he was cresting toward the peak of his influence, coming off a third consecutive NBA championship. But the death of his father, James, in July, as well as dogged and increasingly nasty rumors about his gambling habits, had seemingly forced Jordan up against a sort of breaking point, a midlife crisis, at the tender age of 30. He took a shot from way downtown. He hoped to revive a childhood dream, his father's dream: that he would play professional baseball. To many, it seemed a Plimptonian masquerade.

The story of Jordan's failed attempt to reach the Major Leagues has not been remembered kindly. He was an athlete, *lady*, but not a baseball player. His summer riding the bus with the Birmingham Barons, as a right fielder in the Double-A affiliate of the Chicago White Sox, is mostly characterized by the large crowds and circus-like atmosphere that trailed his every move across the backwaters of baseball, while Jordan's aptitude, as a player, mostly fizzled. In that season in 1994, in 127 games, Jordan made 436 at-bats and collected 88 hits, for an average of just .202. Only 21 of those hits went for extra bases, including three home runs. Though he stole 30 bases and scored 46 runs, he struck out 22.9 percent of the time. Scouts easily assessed his weaknesses and pitchers picked them apart, favoring early inside fastballs followed by soft stuff away. Jordan, at six foot six, had a loopy swing and weak wrist action, resulting in the sort of dribbly contact that only served to enfeeble an athlete of his enormous stature. Many of his hits were simply grounders he beat out with his speed.

It had been 13 years since Jordan had last played competitive baseball. Even then, at Laney High School in North Carolina, he was not really a standout on the diamond. He thought he might walk onto a college team. As a 12-year-old, Jordan had been named the state's youth baseball player of the year. But it was largely on account of his pitching, hurling two no-hitters during the season. In a regional tournament in Georgia, he hit a 265-foot home run that his father, James, would regularly remind him about, even as he was winning MVPs for the Bulls. In 1990, James Jordan suggested to Michael that he probably could make it in the big leagues, following in the footsteps of dual-sport football-baseball players Bo Jackson and De-ion Sanders. "You've got the skills," he told him.

But James Jordan's overzealous confidence in his son's ability to seamlessly transfer his supernatural skills in one sport to another was clearly misguided. Others more quickly (and ruthlessly) confronted the absurdity of his endeavor. BAG IT, MICHAEL! said the cover of the March 14, 1994, edition of *Sports Illustrated*, JORDAN AND THE WHITE SOX ARE EMBARRASSING BASEBALL. Players and coaches quietly implied that Jordan was on a fool's errand. "It's like climbing Mount Everest," the Hall of Famer Dave Winfield said at the time. A "million-to-one shot" was how White Sox general manager Ron Schueler described it. "It's a fascinating experiment," a *New York Times* editorial put it. "As if Albert Einstein had decided to give up physics to try finding a cure for cancer."

There were some obvious protectionist attitudes as the game closed ranks. No outsider should be able to don a pair of cleats and a bag of sunflower seeds and call himself a big leaguer on a lark. "He had better tie his Air Jordans real tight if I pitch to him," the Seattle hurler Randy Johnson sneered. Jordan, ever the competitor, embraced the challenge, working with a hitting instructor for so many hours that his calluses reportedly would tear and his hands bled. Even his staunchest supporters might have conceded that, with every whiff, Jordan seemed to be spiraling toward career immolation. But it may be worth taking a moment to ponder the focal question of his experiment: If Kruk could do it, what exactly was so constraining for Michael Jordan?

In 1932, the British psychologist Sir Frederic Bartlett postulated that hitting a ball in tennis or cricket involved the retrieval of past strokes lodged in memory, which is about true. The ball arrives so quickly that everything from the plane of the swing to the pattern of the step needs to be prepared, if not initiated, before the ball flight begins. To Bartlett, the timing of the unfurling of those memories enabled the actor to appear as if he had "all the time in the world" to do what he wants. But, in fact, timing is relative. Five years earlier, Stanford football coach Glenn "Pop" Warner had commissioned the psychologist Walter Miles to examine reaction times of linemen as they received a snap call. Miles found that the times could be sped up—by as much as 100 milliseconds—if the call was coordinated, and thus anticipated. And if the start of a baseball swing was based on anticipation, so was its climax. In 1954, two Illinois psychologists, Alfred W. Hubbard and Charles N. Seng, were surprised to find that batters actually could not physically track the ball all the way to the point of contact with the bat. That colloquialism "watch the ball hit the bat" is, in fact, utterly futile. But some may be better at maintaining eye contact with the pitch for longer than others.

In 1984, Terry Bahill and Tom LaRitz examined the ball-tracking ability of Brian Harper, a career .295 hitter in 16 Major League seasons. Harper could only track the ball until it was approximately 5.5 feet from home plate, which helps explain why Mariano Rivera and his late-breaking cutter were so nearly unhittable. Novices lost sight of it when it was about 9 feet away. A typical batter will begin his swing at 19 feet from the plate. By this point, he is still only partly through the process of interpreting what is coming. The neural planning for hitting coincides with the swing itself. Good batters, some have found, will focus on the spot where they expected the pitcher to release the ball from his hand, then quickly shift their eyes to the approaching ball after as much as a seventh of a second. Really good batters, after momentarily extracting information about the pitch selection, manage to shift their eyes a second time: to the spot where the ball is expected to be. Two other researchers (Abernethy and Russell) discovered that badminton experts anticipated the shuttlecock's landing position by picking up relevant cues from the opponents' pre-serve movement patterns. This has been confirmed, over and over, in a litany of other sports, reinforcing the anti-Ruth notion that intercepting a moving pitch is less about reaction than it is about prediction.

To grapple with the thorny question of how batters make accurate predictions, then, one of the most instructive pieces of literature for many coaches and players came from the man many considered to be the best at it: Ted Williams. He titled his seminal book *The Science of Hitting*. It was published in 1971, 11 years after he retired with a .344 lifetime average. A short manual, it described his approach at the plate in academic detail. Williams, a former Marine fighter pilot in World War II and Korea, had exceptional vision and a sharp memory; he wrote that he could recall everything about his first 300 home runs, including the pitcher, the count, the pitch itself, and where the ball landed. But at the plate, he was a "guess" hitter, surmising what pitch would be thrown where, depending on the count and the situation on the field. Williams tried to boost his own chances by studying the tendencies of pitchers. This a priori information was more than most were using in the batter's box at the time. His influence would help a generation of new hitters, who now regularly wander around their clubhouses with heads buried in video of opposing pitchers on tablet computers. And yet, no one has been able to replicate Williams' feat of batting .406 in 1941, or even really come close. Furthermore, how would Williams explain that Kruk (and his diet of beer and cigarettes) could hit .320 at the Major League level while Michael Jordan, whose genius at recognizing the subtlest tendencies and flaws of his opponents, beyond just his pure athleticism, turned him into the world's greatest basketball player, could barely crack .200 with the Double-A Birmingham Barons?

This also puzzled Harold Klawans, a Chicago-based neurologist. He knew that, soon after birth and into early adolescence, there are critical stages of brain development. These stages are more susceptible to modification than other periods. The architecture of the brain changes as it matures. New axons sprout and dangle, stretching for more connections to other neurons, and myelin—an insulating sheath that speeds the transmission of messages between the nerve cells—builds up around certain neuronal pathways. And then, at some point, this period of cortical efflorescence ends. The window slams shut. When, in the late 1960s, neurophysiologists David

Hubel and Torsten Wiesel sutured closed one eye of a few eight-day-old kittens, they found that, after they unstitched it three months later, the cats remained blind in one eye, despite both being perfectly healthy. The neurons of the stitched eye had either been assumed by the open one or gradually withered from disuse. The same probably went for the ability to learn skills. Certain motoric abilities, if not acquired at a young age, likewise become increasingly onerous as we mature. Skills that we might learn in adolescence need to be actively maintained to stay sharpened, lest the synaptic connections that enabled those skills atrophy and disappear. Not only does the window close quickly; it has to remain continually propped open. This, Klawans argued, likely accounted for Michael Jordan's baseball struggles. "The sad fact," Klawans wrote, "was that, at age 31, Michael Jordan's brain was just too old to acquire that skill." At that critical, earlier time, he had been shooting jump shots, not swinging at baseballs. By the time he picked up a bat as an adult, it was too late.

And then something strange happened at the plate for Michael Jordan: He got better.

In August, the last full month of the season, he batted .380 and hit two of his three home runs. He was invited to play in the Arizona Fall League, a showcase for top prospects, and batted .255. The footage shows a more confident swing, more life, better contact. Had Michael Jordan turned the corner to finally find his path to the Majors? OK, probably not. Eventually, he would likely reach a point where he would stop meaningfully improving, and that plateau would remain well short of the prospects who were eight or ten years younger. But Jordan had progressed so much in such a short period that the *Sports Illustrated* writer Steve Wulf tried to follow up his "Bag It" story with something of a mea culpa (the magazine decided not to print it). By the time the 1995 Spring Training was supposed to begin, the baseball strike that had curtailed the 1994 Major League season was still under way, and Jordan did not want to wait around to see when it might eventually lift. He bolted back to the Bulls and won three more NBA titles.

"The nerve paths are something fixed, ended, immutable," Santiago Ramón y Cajal once wrote of adult brains. "Everything may die, nothing may be regenerated." But like so much in the embattled history of motor science, this idea, once delivered with such confidence, has failed to live up to the scrutiny. Cajal's "harsh decree" in the early part of the twentieth century and Klawans' "critical window" theory at the end have each proven to be much too unyielding. The evidence for neurogenesis and neuroplasticity, the new growth and rearrangement of neurons, now casts the brain in a more forgiving and fertile light. The cerebral cortex, which is the brain's wrinkly outermost layer, contains and is itself a shifting and dynamic landscape, like beans within a beanbag. When researchers at Stanford fitted a group of barn owls with prismatic glasses to distort their vision, such that a mouse placed on the other side of the room would appear in a different location than its squeals, the owls slowly learned to adapt. The optic tectum in their brains morphed to reflect that. When a neuroscientist at the University of California, San Francisco, stitched together the index and middle fingers of an adult monkey, the mental representation of its hand changed to reflect four digits instead of five, the two fingers having fused into one. And when Eleanor Maguire, at University College London (UCL), examined the brains of London cab drivers in the late 2000s, she found that they could express drastic modifications as well. The drivers are required to pass a rigorous set of exams, called the Knowledge, which encompasses a mind-boggling 25,000 streets and 20,000 landmarks along 320 different routes. Most drivers study for several years before they are up to the test. Maguire looked within the hippocampi—hook-shaped brain structures known to be involved in memory and spatial navigation—of 79 trainees and found that the more time the drivers had spent behind the wheel, the more change she saw. In fact, the posterior hippocampi appeared physically larger than those of nondrivers. They were larger than London bus drivers'. She tested drivers before they started and four years later and found that that region of the hippocampus had gotten significantly larger in those who had continued to drive. There was no growth among those who dropped out.

"The brain-matter is plastic," the psychologist William James wrote. "The hemispherical cortex shows itself to be so peculiarly susceptible" to impressions. Those impressions are made through the senses. Training, whether by driving the streets of London, studying rigorously for a medical exam, or fingering the strings of a violin, produces tangible effects on the organization of regions of the brain most responsible for that behavior. The hippocampi of the cabbies reflected the intense requirements on spatial navigation, as though their brain was a quadriceps muscle being bolstered by squats. And yet—what is the brain region responsible for hitting a baseball?

You might suggest, as many have, that good hitting is mainly owing to good vision. And baseball players do generally have sharper eyesight than the normal population. Williams, for instance, was said to have 20/10 vision, birthing a myth that he could follow the seams of the baseball as it came toward him (Williams denied it). Michael Jordan's vision is said to also be superb; Al Michaels, the broadcaster, once remarked after golfing with Jordan that he seemed to observe the world in "4-D." But in 1993, a team of ophthalmologists and optometrists conducted a study on the eyesight of the Los Angeles Dodgers and found that the mean visual acuity was roughly 20/12. The theoretical limit of human vision is 20/8. It was not as though half the Dodgers were on the All-Star team, and not everyone with 20/12 vision is a Major Leaguer. In fact, a 2010 study found that skilled cricket batsmen could still successfully hit live bowling while wearing contact lenses that blurred their visual acuity to 6/49, a level that approaches legal blindness. An ophthalmologist who examined Ruth toward the end of his life reportedly found something that the Columbia researchers possibly never noticed: The slugger was congenitally amblyopic. He appeared to be legally blind in the left eye.*

One of the authors of the Dodgers study, Daniel Laby, has continued to consult with baseball franchises about the role that vision plays in hitting. When I met him in his office one afternoon, he was wearing a Cleveland Indians cap and a Chicago Cubs jacket, both to indicate his relationship with the clubs and to reinforce his image of objectivity as they were playing each other in the World Series. He told me about one of his favorite specimens for study—a favorite specimen for a lot of things—the great, oafish, sui generis Boston Red Sox slugger Manny Ramirez. Even during his prime, Ramirez did not have the best vision on the team, Laby said. He apparently had something better. Laby devised a training exercise involving a hollow plastic ring the size of a Frisbee. The ring was placed through

---

* The exact time of the development of Ruth's disorder (diagnosed as amblyopia ex anopsia) has never been conclusively verified.

a Wiffle Ball, which the player is supposed to catch—grabbing only the ball—after it is tossed from the other end of the room. The drill tests both visual and motor planning—following the path of the ball with your eyes and closing your hand at the right moment to intercept it. To add another wrinkle of difficulty, Laby positioned four Wiffle Balls along the ring and painted them different colors. He would yell the color he wanted you to catch as it came toward you. The first time Manny Ramirez tried this he was perfect each time. "Doc," he said. "Too easy." Fortunately, Laby had a third ring that intersected four baseballs. Instead of different colors, though, these balls were painted with the subtle lines of a different spin pattern. As the ring traveled halfway across the room, Laby would call out a pitch—"Fastball!"—and Ramirez would have to grab the corresponding baseball. Laby had never tried this with a professional athlete, but it was clear that Ramirez could "see the ball" better than even his teammates with greater eyesight. "Manny didn't have the best vision on the team, but he had the best package," Laby said. "He was just able to hit the ball." The ring drill became a staple of Ramirez's warmup routine.

Michael Jordan might have had the vision, the hand speed, the coordination, even the intuitive know-how to better predict what a pitcher could be imminently unloading. In fact, this inimitable athletic concoction is probably what enabled him to bat even .200 at the Double-A level, essentially having come off the street—an average that probably should be marveled at, not ridiculed.* But another

---

* By comparison, another superb athlete, Tim Tebow, the former Heisman Memorial Trophy winner and NFL quarterback, has currently been attempting essentially the same experiment, with even less success. In 145 professional games, Tebow had yet to advance beyond the Single-A level for the New York Mets.

ingredient was still missing from the "package" that Michael Jordan brought daily to the plate with those blood-soaked hands and that picturesque frame. The ingredient, of course, resides beneath the helmet.

I listened as Laby guided me through the assorted "levels" that a tiny photon of light passes through to trigger a response from the motor system, from the watery tear film to the retinal ganglion cells to the optic chiasm to the visual processing areas in the rear. There are only about a million optic nerve fibers, yet almost half the cortex is involved in processing the signals they transmit. For a signal to correspond with a movement, it has to be imbued with goals, memories and emotions, all arriving from other portions of the brain. The tiny photon of light is now heaped with the baggage of the mind. The tiny photon is not a photon anymore, but an impulse that has ping-ponged from passive perception into voluntary action. It is the coin that slides into the jukebox. The cost is approximately 200 milliseconds, from the detection of a visual impulse to the response of the musculature.

A fastball traveling at 95 miles per hour takes about 400 milliseconds to reach home plate from 60 feet, 6 inches away. That does not account for the length of a pitcher's stride, or the deception pitchers employ with their delivery, or the fact that 37 pitchers in 2016 *averaged* more than 95 miles per hour with their fastballs. In the amount of time it takes the pitch to reach the plate, the physical limitations imposed just on our bodies have already sliced our available response time in half. The resulting amount we have to actually gauge the pitch is almost twice as fast as an eye blink. It is slightly slower than the duration of one rotation of a helicopter rotor blade.

But in the time it takes to read this word, the ball will have sailed past. It should not seem a wonder, then, that it has been more than 75 years since a Major League Baseball player batted .400. It should seem a wonder that our brains enable us to ever hit the ball at all.

Yogi Berra famously asked, "How can you think and hit at the same time?" For Jason Sherwin, the answer arrived early. Growing up, he often struggled at the plate. He had a hitting coach and attended hitting camps, but pitching came more naturally than batting. "He thought too much about it," his mother, Judith, said. "He was like that in basketball. By the time he figured out to try to make a basket, they knocked it out of his hands."

Off the field, he was a bright student and a talented musician. In the East Rogers Park neighborhood where they lived, two blocks from the lake on Chicago's North Side, he took piano lessons every Sunday for a decade. For almost two decades, he also studied the Talmud with the Orthodox rabbi who performed his bar mitzvah service. After he left Chicago, they continued to study over Skype. It pleased Jason's father, Byron, a conservative rabbi and renowned Jewish scholar, that Jason maintained such interest in his religion. But it pleased his mother that he maintained such interest in hers: baseball. She was the lifelong Cubs fan, the one with the season tickets in Section 220, Row 7, at Wrigley Field, and the one who coached his youth teams. She introduced him to Bob Prokopowicz, hitting instructor of a local baseball academy, who first talked to Jason about the mental side of hitting, well before Jason would come to devote his career to it. She enjoyed the time her husband took him to the airport to pick up "a cardinal." Jason, then 12, expected to find the St.

Louis Cardinals shortstop Ozzie Smith. Instead, it was Cardinal Joseph Bernardin, one of Byron's longtime friends.

So when she heard that Jason had begun a project at Columbia to study the brain patterns of baseball players, she was not particularly surprised. It combined two of his lifelong interests: baseball and science. At age eight, Jason entered a science fair with a model for how Shaquille O'Neal could produce the force to bring down a basketball hoop. At nine, his parents gave him a book by Michio Kaku. A few years later, he could not put down *The Physics of Baseball*. At the University of Chicago, where he completed his undergrad, he took one course in neuroscience and majored in physics (his other major was music). But there would come a shift toward applied science, propelled by an interest in outer space and a household that loved *Star Trek*. He looked at graduate schools that allowed him to pursue aeronautics and, as he put it, "build rocket ships." Judith saw a more essential impulse. "He wasn't interested in theory," she said. "He wanted to do things." His grandmother once asked him as a teenager what it was he wanted to do. Jason replied, "I want to do something to change the world."

Jason's whimsicality somehow married nicely with the quiet desk mate he got to know in Paul Sajda's lab at Columbia. Jordan, a reserved and steady worker, shared with Jason at least one thing in common: Neither exactly knew what they were doing there or what they wanted to do. Jason wanted to build; Jordan wanted to understand. Jason, now 34, is tall and lean, with tropical blue eyes, wavy auburn hair and loosely maintained bearded scruff. Jordan looks as though he treats facial hair like an invading army. They dressed differently—Jordan: khakis; Jason: half-buttoned linen shirts. Jason's clothing can sometimes appear to be hung from his body, rather than worn. While Jason spent his spare time as the front man for a rock band called The Conditionals, Jordan never created a Facebook page. Jordan has a 1.6 handicap in golf, a sport for which Jason has no patience. They also speak at a different volume and pace. Jason never seemed at a loss for things to say, a trait he thinks he developed not from his mother (a trial lawyer) but his father, who led services during the High Holidays. Jordan, whose parents divorced, gets nervous in front of crowds. He, at least, could recognize immediately how different they were from each other. Jason did not seem to even notice it. "Do you think we're like Oscar and Felix?" he asked me once. He was ironing jeans in a Virginia hotel room as Jordan nodded off to sleep after a long day at a convention. It was 9:36 at night.

But they were not businessmen, and neither of them had ever taken a course in business or marketing. Frankly, they were latecomers to neuroscience, too. Jason had done his entire graduate schooling in aerospace engineering at Georgia Tech, which he hoped could position him for a job with NASA. But it was a book called *On Intelligence* by Jeff Hawkins, the creator of the PalmPilot, that for some reason stuck with him. "I just tore through it." He said: "I can't even recall how I ran into it. A lot of things started making sense, in terms of observations. The basic idea was that our nervous system was a prediction-generating machine. A lot of these kinds of things started clicking for me." His thesis became a meditation on situational awareness—in the context of soldiers in the Iraq War—but he was the only one of 80 candidates to use the human brain as a framework. The committee members at Georgia Tech did not know quite what to make of it. But Jason was invigorated. "It laid the groundwork for a bigger idea," he said, involving decision-making and strategizing. It happened to fit what Sajda was looking for. "The soft items on my résumé probably grabbed him more, rather than my exact skills in neuroscience," Jason said. "Because I didn't have any." He added: "He took a gamble on me."

Around the same time, Jordan was also moving into Sajda's lab, a ground-level enclave with an underperforming air-conditioning system. For the past three years, he had been programming the large-scale analysis of brain data sets with up to 1,000 subjects for studies on Alzheimer's and certain vascular conditions associated with aging. As an engineer, he knew how to code the software to analyze the data. But he also grew close with a professor named Truman Brown, who, as a rare physicist in the radiology department at Columbia, has cleverly advanced several aspects of our usage and understanding of magnetic resonance imaging (MRI). Functional MRI (fMRI), born in 1992, is a technique that records hemodynamic changes in blood flow in the brain. As neurons receive signals, they use oxygen, triggering a rush of blood to replenish supply. Deoxygenated blood is paramagnetic; it creates enough of a magnetic field distortion that scientists can then map it. Brown hired Jordan as a teacher's assistant. Under his tutelage, Jordan became, as Sajda put it, "a MacGyver." "He can put things together and do analysis really quickly and thoroughly," he added. Then Brown left. Jordan was in the middle of a project to interface motion correction—movement in the scanner is detrimental to the image, which is why motor studies have historically been so difficult to pull off using fMRI. Sajda, who had been collaborating with Brown for almost a decade on methods of combining EEG with fMRI, took him on. "I was the only person who really knew how it worked," Jordan said of fMRI. "I became kind of a liaison to run all our experiments up there" at the scanner on the medical campus.

In the fall of 2010, Jason joined a desk with Jordan. He lamented the state of the Cubs while Jordan complained about the Yankees. Otherwise, they had little to do with each other. Jason decided to study musicians, and he pressed Sajda for more opportunities for applied research. For one experiment, he gathered expert cellists and had them listen to a piece of music while in a dark room with EEG. He would faintly alter the music, ever so slightly, to move the key from G to G-flat. As the music continued in G-flat, the harmony became normalized again. It was not rough on the ears. The difference was almost imperceptible. Only, something interesting occurred in the brains of the experts at the time of the key change. Their motor cortex activity became significantly amplified. Their brains responded to the change. The brains of novices did not reflect this. Both parties sat in the dark, quietly, and nobody actually moved their arm. "But mentally," Jason said of the cellists, "they wanted to." Buoyed by the results, Jason presented to the lab for feedback. The difference of the brain responses within the neural circuitry of experts and nonexperts made Sajda encouraged about looking further, but it was Jordan who took the most interest, pulling Jason aside afterward. "Do you think we might see this with athletes?" he asked.

Jordan had been taking golf lessons during the afternoon with Columbia's golf coach. His first idea was to use the EEG to study golf. Several neuroscientists, including John Milton at the Claremont Colleges, had examined expertise in putting and concentration and discovered exciting differences in the organization of neural networks that might allow for better focus on the green. Jordan wanted to know how different brains responded when they watched somebody else's golf swing. Jason did not. "Let's do baseball," he kept saying. Jordan would soon learn something his friend's mother had known about her son for a long time. "Jason's the most stubborn person I know," Judith said. "We settled on baseball," Jordan said.

They tested six subjects with little to no baseball experience, as something like a pilot study. "It was kind of a pet project," Jordan said. "Like, wouldn't this be cool if it worked." He programmed an .avi movie file onto a Dell Precision laptop, while Jason tore through his *Physics of Baseball* book for equations for the velocity and movements of a fastball, slider and curveball. The ball was represented by a green dot against a gray fog in a computer simulation. The dot increased in size as the "pitch" approached, offering the illusion of depth. The participants, like the rookie ball team, were asked to identify the type of pitch as quickly as possible after the prompt, or "release." They did this by tapping a key on the laptop's keyboard if the pitch they received matched the one they were told to expect. Many times, it was hard to tell. I tried the simulation in early 2015, when it was still roughly the same iteration. I visited the deCervo headquarters, which consisted of a desk in a coworking space in SoHo (they have since moved to a desk in an office building in a different part of SoHo), and Jordan hooked me into the EEG, starting with the amplifier, a device no bigger than a radio scanner, which was held in place at the back of my neck by an elastic bandana.

The equipment, made by a company called Advanced Brain Monitoring, cost $14,000 and could be transported in a briefcase, which is one of the reasons that make EEG (short for electroen-cephalogram) use so appealing. As a measure of cortical activity, electroencephalography is most commonly associated with assess-ments of "brain waves," or bursts of electrical responses within large groups of neurons. There are said to be five types of waves that fall along a spectrum, as measured in hertz (Hz):

- 0.5–4.0 Hz: delta. Your brain's typical activity level when in a deep sleep.
- 4.0–8.0 Hz: theta. A half-sleep state.
- 8.0–13.0 Hz: alpha. Awake, relaxed, focused.
- 14.0–26.0 Hz: beta. Deep in concentration.
- 26.0–100 Hz: gamma. Seen in brief, fleeting moments of infor-mation processing.

In the 1960s and '70s, scientists learned to use these EEG read-ings as a technique for helping patients manage the activity within their own skulls, through a process called "neurofeedback" or "bio-feedback." In 1972, John Lennon and Yoko Ono, sitting cross-legged on a forest-green carpet, tested EEG on *The Mike Douglas Show*. "I hope my alpha's all right," Lennon joked. Its growth in popularity was often associated with yoga, Transcendental Meditation, mind-fulness training, Eastern medicine—means of settling into a more healthful, mindful state, say, from reducing your beta moments into thetas. EEG could show you where your brain's challenge spots are—if, for instance, your activity pattern was showing too much theta in the frontal lobe, you might have difficulty focusing in high-pressure situations, and through training, you might be able to fix that. That's the evergreen promise of neurofeedback, which remains in widespread usage today, including in sports settings. In the run-up to the Super Bowl in 2014, the Seattle Seahawks started using neurofeedback through EEG to supplement their already progressive supply of yoga and meditation routines, as did Kerri Walsh Jennings, the Olympic beach volleyball star.*

This is not how Jason and Jordan use EEG, however. They use EEG as a recording device to eavesdrop directly on the natural exchanges happening between populations of nerve cells and their thousands of connections simultaneously. By contrast, fMRI is only capable of measuring blood flow as a result of neural activity, not the activity itself. Because it takes only a fraction of a second for a neuron to send one of its action potentials (the impulse used to communicate information) to another neuron, the device recording these exchanges needs to work at that time scale. Jason and Jordan's device could take 2,048 snapshots per second. One second of this EEG signal can look like Himalayan peaks. The EEG gathers the information through electrodes placed along nine different points around the head, held in place by a clear lattice of pliable plastic— the luminescent "swim cap." "Players don't want to wear something that looks like a Hydra on their heads, with all these wires going everywhere," Jordan said. "They want small and sleek." Once the cap was connected, Jordan flipped on the game. Jason would later joke to me that "Atari was light-years ahead of what we were doing." He was not wrong. It was surprisingly rudimentary, but also surprisingly engaging. The dot moves, depending on the pitch, and

---

* Proof of the benefits of neurofeedback through EEG remains almost entirely anecdotal.

enlarges as it approaches. The mean pitch speed was just 78 miles per hour, but some curveballs were slower and some fastballs were faster. Sliders moved side to side and the curveballs dived down at a steep angle, or what pitchers call a "12-to-6" break. Fastballs remained straight (to me, actually, they appeared to tail upward, a common illusion). As the pitch loads (that is, the windup), the prompt will indicate which pitch to expect: fastball, curveball or slider. The prompt, of course, could be lying—it is your task to "swing" if that pitch was delivered, or "lay off" if it was a different pitch. Many fastballs looked like sliders and many sliders looked like curveballs. And 78 miles per hour is batting practice speed for college players. But for a novice, even someone who played baseball growing up, the velocity was jarring at first. In my first trial of 90 pitches, I accurately decided whether to swing or not swing only 52.26 percent of the time. I was basically guessing.

I did improve in my second trial, but the game, at that point, was not necessarily about improvement. It was about assessment. After the exercise, Jordan showed me my sliding window logistic regression (the milliseconds when my brain firmly decided to swing or not); a side view of how far along the pitch was in its trajectory before I decided; and an analytic called "Swing vs. Take," which determined my accuracy when I made my decision between, say, 500 milliseconds and 600 milliseconds (80 percent). An experienced baseball player, they have found, will expect to have 80 to 90 percent accuracy 400 milliseconds after the ball is released. Of course, the longer you have to get a look at the incoming pitch, the better your chance of reading its spin and location. But real hitters do not have that sort of luxury. My success rate at 500 milliseconds, while reasonably accurate, would nonetheless leave me standing at the plate with the bat on my shoulder.

In the initial study with six participants, Jason and Jordan found something else of interest after they applied a technique called source localization. When the subjects missed the pitch—tapping when they should not have or not tapping when they should have—a strong burst of activity appeared in the left prefrontal cortex, like water bubbling up from a spring. It was located in Brodmann Area 10, a spot associated with working and prospective memory. The activity they saw was occurring after the pitch had passed, representing perhaps a momentary flutter of evaluation, a tinge of regret. The brain was telling the researchers that they were seeing pitch recognition unfold. "It was like, 'Aha, now we've got something we can measure,'" Jason said. "And we've got something we can fix." They published the results in *Frontiers in Neuroscience* and it quickly amassed interest. "It was not only that we did an OK job on the science part of things, but also that other people that don't know this field thought it was worthwhile and interesting," Jason said. "That was what blew me away." Jordan, as usual, remained more cautiously optimistic. "It ended up working shockingly well," he said of the study.

Their next step was testing experts. Columbia's baseball team had been one of the best in the Ivy League, and with Walker's help, Jason and Jordan secured some access for experimentation. Sajda encouraged them to probe further than EEG recordings. His area of expertise allowed them to use the EEG in combination with fMRI, which provided the spatial resolution to give a clearer picture of where the brain activity was being generated. The EEG could still give the temporal specifics. Twenty ballplayers went into the scanner and 20 different brain responses came out. Their cortical regions lit up like fireworks against a night sky. A colorful picture of the brain inside the helmet of a hitter began to form. The researchers did what anyone might do with that sort of information at their fingertips. They used that picture to compare against the brains of non-players like themselves.

Utilizing the EEG and fMRI recordings, they could now chart when and where specific activations occurred within the life span of the pitch. When the experts decided to swing, the supplementary motor area activated at 250 milliseconds; the paracingulate gyrus at 300; the hippocampus and middle temporal lobe at 375; the posterior cingulate at 425; and the supramarginal gyrus at 525. The right superior frontal gyrus responded faster for pitches that were correctly let pass (275 milliseconds) than for pitches at which the batter swung (300 milliseconds). When the experts decided not to swing, there was higher activation in the parahippocampal gyrus and the paracingulate gyrus—early visual processing areas said to be used for visual prediction. *Less about reaction than it is about prediction.*

When the experts decided not to swing, that inhibitory urge also correlated with activation in the supplementary motor area (SMA), a dense blob of cortex behind where the logo might be on a batting helmet. Among other things, the SMA is implicated in movement initiation. Studies on Parkinson's patients often reveal abnormal connectivity between the SMA and other brain regions, like the thalamus. Parkinson's patients also typically have trouble with timing. Neurons in the SMA need a strong stimulus current to evoke a movement, but unlike the primary motor cortex, they can also stop one in progress. It has been shown that the SMA activates when people are simply imagining a movement, as though it plays a suppressive role in that, too. "They're kind of on a hairpin trigger," Sajda said of the baseball hitters. "They're planning to go, but the good player has the enhanced activity that stops them." The researchers then saw a different brain region involved when the hitters did decide to swing.

The fusiform gyrus is a narrow strip along the brain's underbelly almost directly opposite the SMA. But its role is visual recognition, particularly of faces. A person with prosopagnosia, or face blindness, often has an impairment of his fusiform gyrus. More recently, it has found its way into the perceptual literature on expertise. When an automobile enthusiast sees a souped-up Alfa Romeo Spider turn the corner, his fusiform gyrus lights up in a different fashion than if he simply saw a standard Alfa Romeo Spider. The same thing happens when a radiologist examines scans; when a young card collector sees the picture of a Pokémon character; when a birder spots a distinctive plumage; and when a chess master surveys a board. A portion of the fusiform gyrus is said to be involved in identifying words. Now it was active when a baseball expert selects a pitch. Though it is tempting, and very often misleading, to associate one brain area with one function, there seemed to be a reason the fusiform gyrus kept coming up again and again in the literature. Most of us are experts in recognizing faces. Most of us are experts in reading words. The notion that the same neural resources might be recruited to help an expert hitter recognize his pitch as his pitch was not altogether unbelievable.

The engagement of the limbic system commingled with the

motor inhibition signals in the SMA and the recognition read-outs of the fusiform gyrus altogether painted a compelling picture. That all this was bursting forth at different times within a matter of milliseconds from pitch release to swing was even more compelling. The swing, as Jason likes to say, is only the tip of the neural iceberg. Most of a movement is precipitated beneath the surface. The link between perception and action, seeing and doing, is more than just reactive it is also shaped by experience and guided by prediction. It involves both memory and anticipation. Most cognitive neuroscientists believe the brain interacts with the world around it by forming representations, which map the body and the environment from the sensory information that is available. Representations are little more than patterns of neural activity, but it is said that neurons that fire together strengthen their connection together. The reason we can improve our tennis strokes over time is often unfairly mislabeled "muscle memory." Of course, muscles don't have memory. They act on the whim of the brain, which reshapes and reorganizes itself by bolstering the connections required to more easily and efficiently produce specific commands to the limbs.

Expertise in a sport is regularly referred to as "seeing the field differently" than others. It is true that our experience does seem to change our perception, which, in turn, influences our actions. We have more or less guessed this from behavior. Now, as Jordan says, there are ways to peek beneath the hood. If certain activity patterns can be discerned from the outside, the correlates of skill might be analyzed like the configuration of clouds in the sky. See enough clouds and you can forecast the intensity of the storm blowing up on the horizon. As they sailed across the darkened cortex, Jason and Jordan began to recognize the dim contours of an impending gale.

The results were telling them the experts' brains treated pitches differently than the novices'. Their intuition was telling them they could start a business informing teams about the differences. As they began the EEG testing of more players—first from Columbia, then from places like Bradley, Brown and the University of Illinois—they saw neural markers distinguishing hitters even from within the same lineup. A leadoff hitter with a knack for recognizing pitches early. A cleanup hitter with drifting focus before the pitch is released. A top recruit who responds to fastballs much more decisively than to curveballs. After one visit, Dominguez, the Bradley coach, was spellbound. "You guys were here for an hour," he told them. "You didn't even see them play. And you could essentially give us the same stats we see on the field." In fact, they were better. They were not suffused with inherent biases or rammed into post hoc narratives intended to project future performance. They were straightforward, from the source, as effective as a radar gun.

Jason and Jordan began to envision their brain data appearing on scoreboards right next to a hitter's batting average and home run totals. They had a quixotic vision of what EEG could do for scouting, as a baseline method for determining if a prospect had the visuomotor traits to cut it in the big leagues. Jason imagined a not-so-distant future where a dogged scout trudged through slum villages in India, metal briefcase in hand. The scout might stumble upon a boy with no baseball training but a peculiar capacity for intercepting falling acorns with a stick. Like the young Indian pitchers in the Disney movie *Million Dollar Arm*, talent could be identified far away from the field. "We're helping baseball teams find the million-dollar brain," Jason said.

One hot and sun-splashed spring afternoon in Arizona, which could really be any spring afternoon in Arizona, Jason and Jordan settled into seats on a small set of bleachers between home plate and the home dugout, a view that offered them a wide vantage point of both the pitcher and the batter in an exhibition between Minor League clubs. It was extended Spring Training in 2015. Already they had spent the morning examining the players' neural responses with their novel simulations. They wanted to see the game in real time. After a while, a hitting instructor from the Oakland Athletics hopped into a seat next to them. Together they watched a batter. As the pitch arrived, the batter lunged out slightly over the plate. "You see that?" the coach asked. "That means he's not seeing the ball well."

Jason and Jordan had not noticed it. But the coach said it was something he had spent the last decade training his eyes to detect. The conversation then veered from what the batter was doing to what he was not doing—seeing the ball all the way in, keeping his head still, and keeping his eyes focused as the pitch approached. Instead, because he was having trouble recognizing the pitches, his body induced a response (lunging) to better evaluate the trajectory or direction of the ball. Perception influences action, and action also influences perception. The psychologist James J. Gibson said this another way: "We must perceive in order to move, but we also must move in order to perceive." This batter had to move in order to perceive. But this took up precious time and it took him out of position. Jason offered a quick evaluation: The batter's perceptual mechanism—his rapid perceptual decision-making—wasn't accurate enough for him to know where the ball was going or whether he was going to hit it. So he adjusted his approach, something hitters do all the time. If they are slumping, or having difficulty seeing the ball, they might open their stance or quicken their stride. Coaches might advise them to try to improve their focus, settling on a specific target during the pitcher's windup, as George Brett (his hat) and Steve Garvey (his face) said they would do. Or they might choke up for better bat control, a tactic routinely employed by Tony Gwynn.

Coaches have always had theories about hitters' tendencies when they were slumping, and savvy pitchers can often pick up on these tendencies as well. The game has existed that way for 170 years. But now, the guys told the Oakland coach, there is data that doesn't rely on guesswork, based on cues that don't sit out there for the pitcher to detect: *brain data*. With this, he could have a more concrete understanding of the issue and how to change it. Maybe the coaches might throw a struggling batter more off-speed pitches in batting practice, or lengthen his swing. A hitter might improve through video simulation, or his brain might self-correct on its own, armed with better feedback. Instead of guessing at a solution to a problem—lunging at the ball or choking up—a hitter might know how best to fix it. As the deCervo founders discussed this, they could see the instructor's eyes illuminate with the possibility of such a window into a hitter's fragile mind. "It was kind of like how I explain to my grandmother how Tinder works," Jason said. The art of dating—not unlike the art of hitting—had been boiled down into an algorithm where it was then reconstituted into something that could be improved with the swipe of a finger. "That was not recognized as a possibility before," he said.

By now, they had logged a full range of differences between experts and novices in elegant detail. They had found the timing differences with EEG and the spatial differences with fMRI; they had even used a technique called a "resting-state scan" to locate differences in the connectivity between the critical brain regions. They felt they finally had an answer to the question: "Can you make my team better?" Yes, they could say. In fact, we think you have been looking at hitting all wrong.

Jason and Jordan did not care so much how the hitters developed their talent. They cared about describing it, in digestible data bites, to teams hungry for that information. But they had already begun to tread, perhaps even unwittingly, into a realm once reserved for poets and philosophers. Their studies could have cited inquisitors reaching back to antiquity. The effort to understand the role of the slugger's SMA was just another version of the cerebellum that Erasistratus studied at Alexandria in the third century BCE. Galen wrote treatises on the cranial nerves before becoming the personal physician to Severus. The ventricles fascinated Da Vinci, who cast them out of hot wax. Descartes believed in the ethereal spirits rumbling through the nerves like steam through a pipe. Isaac Newton saw no spirits; he believed the nerves transmitted information by vibration. Darwin wrote of the size of man's brain, while William James wrote of its mechanistic simplicity. Thomas Edison wrote of little people arranging memories in the hippocampus. Not every immortal thinker produced immortal thoughts about the brain and its functional organization.

I found a scientist, Daniel Wolpert, who has investigated the source of our differences down to the composition of the neural signals that fire. Neurons are like snowflakes; each has individual characteristics. Individually, neurons are almost impossible to classify. But grouped together they produce thoughts, actions, and everything in between. When they fire their action potentials, they can sound like tennis balls pounded against the strings of a racket. *Pop pop pop*. They can sound like batting practice.

2.

THE MOVEMENT CHAUVINIST

WHY WE HAVE A BRAIN

Whenever Daniel Wolpert is asked to give an interview, speak at a conference, or entertain a guest at the Elizabethan formal hall at Cambridge's Trinity College, where he gets to dine as a fellow beneath portraits of other alums, such as Isaac Newton, Lord Tennyson, and Francis Bacon, he typically opens with the same story. It is a personal story, intended to rationalize how a brilliant scholar—trained in mathematics, medical degree from Oxford, fellowship in cognitive science at MIT—could be reduced to experimental work involving finger taps and small, constrained reaching movements. He even gave it at a 2011 TED Talk, now viewed more than 1.7 million times. The story lays out the case for "the real reason for brains." They exist, he argues, to produce movement. "You may think we have one to perceive the world or think," he says. "And that's completely wrong." In Wolpert's view, there is no other explanation. If you don't believe him, you could consider the humble sea squirt, which is born looking almost like a tadpole, with a brain, a nervous system, and even an eye. Throughout its juvenile life, which is only a few days, it swims around the ocean, siphoning in planktonic food particles from the seawater it ingests. At some point, it decides to attach itself to a hunk of coral, at which time the sea squirt summarily digests its own brain and never moves again.

There are other living things on this earth that don't voluntarily move: trees, plants, fungi, sponges and the like. None of these organisms contain a brain. Movement is one of two ways we as humans have to affect the world around us, Wolpert says. The other is by sweating. Everything else is done through contractions of muscles. You communicate by moving the tongue and throat muscles that produce sound, or by writing, tweeting, or gesturing in sign language. Wolpert argues that the only reason for senses, emotion or even memories is to drive or suppress our future movements. The reason we can perceive such things as the color of a bird or the sound of a horn is so it can influence the way we are going to move later in life. Even our memories are shaped by movement. Experiences from very early in life are more easily forgotten than the ones just had. That is because they are less essential for action. I move, therefore I think.

Wolpert is a self-described "movement chauvinist." From his lab on the fourth floor of the Department of Engineering building at the University of Cambridge, he is dedicated to answering the very basic questions of how the brain is able to produce consistent, complex, adaptable movements, again and again, at lightning speeds, with only a minor delay (we'll get to this last point later). Yet he still affectionately refers to motor control, his adopted field, as the "poor man of neuroscience." When he started in the late 1980s, the research did have a reputation for being synonymous with long-winded equations, maladroit equipment and lugubrious tasks. To be fair, the equipment has since improved. Motor control research might not be dripping with the appeal of stem cell research or connectivity or optogenetics. But the reason Wolpert has remained loyal to the field is that he always understood motor control to be the output of almost all the brain's vast array of resources, elided into one purpose: to move the muscles. He cares less about the thinking required to excel at chess than the dexterity required to manipulate a chess piece. His lab, which doubles as a machine learning department, is ultimately interested in designing robots that can accurately mimic the movements of humans. His computations, and those of many others, have already enabled scientists to be able to use signals from the motor cortex to assist robotic prostheses, giving renewed mobility to impaired patients, by incorporating what are called "brain-machine interfaces" (BMI). This is an area that might usher in the future of movement.

At this point in time, though, the gulf between purely robotic actions and purely human actions is astoundingly wide. It can be surprising when you consider how good artificial intelligence has gotten at replicating and building upon strictly cognitive tasks like, say, strategizing in chess or winning on *Jeopardy!* A five-year-old child can display a fluidity of dexterity and range of motion that the most sophisticated computers cannot begin to replicate. Find videos of robots attempting to open a closed door and you are almost guaranteed a good laugh. There is not yet any model for movement from which programmers can steal to encode in machines. That's not a knock on the machinists. It is a reminder to step back and appreciate how we move. "When I watch people do skilled things, it's amazing," Wolpert told me when I visited him in January 2017.

Indeed, when you stop to appreciate the kinematics of our bodies, the wonders pile up. A simple act like taking a step involves a litany of synchronous and coordinated synergistic and antagonistic muscle contractions. Muscles can only pull; skeletal levers are required for propulsion. The glutes and hamstrings utilize the hip joint to move the leg backward. The quadriceps use the knee joint to straighten the leg. The calf muscles use the ankle joint to swing the foot down and back. When the foot is planted, and these muscles contract, it creates the backward push of the leg that enables the rest of our body to move forward. Muscles also can't actively lengthen, so they need to be reset by so-called antagonists, or flexor muscles, on the opposite side of the joint. The antagonist for the calf muscle is the tibialis anterior, which runs along the shin to the top of the instep. The iliopsoas muscles stretch the glutes. The hamstrings stretch the quads. The foot must be off the ground during these movements; otherwise, we would always end up back where we started. Thankfully, nature gave us a second leg for balance.

The rhythmicity of locomotion allows us to forget about its intricacy, and yet the ability to walk is produced by the spinal cord, automatically, without even the need for intervention from the brain, which has bigger things to worry about. At any point when a more complex, voluntary movement is desired—be it a shrug of the shoulders, a flick of the wrist, or the swing of a baseball bat—the brain gets involved, dispensing its signals. These signals, in a nutshell, get consolidated in the cortex, descend down the corticospinal tract from the brain to the spinal cord, and then interface with a peripheral nerve, a portage to the fibers that innervate the muscle. The motoric response to a command is fast, but not instantaneous. And the bigger and faster the movement, the bigger the command needed to trigger it.

This, inevitably, leads to some errors in transmission. The human motor system is not perfect. Every signal, as it travels along the neural pathway to the alpha motor neurons in the spine, and out through the peripheral nerves before making the synaptic connection with the muscle fiber, triggering its contraction, is subject to some random fluctuations. Neurons may fire at irregular intervals, even within the brain, the motor system's most important component. But other mechanisms along the path to movement are also susceptible to unpredictable perturbation. Researchers call this "noise." Not in the auditory sense, but like the static that afflicts a radio signal, crackling an otherwise clear broadcast. It's jitter. Noise is a constant in our biological makeup. The small variances in our signals can yield meaningful consequences in our behavior, to the point that it can be said that no two movements are exactly alike. Why is it so difficult to reproduce the same throw in a simple game like darts? Why is there always some variability when we try writing our signatures the same way every time? Why has no one in the NBA ever made 100 percent of his free throws? Motoric variability is a fact of life. There is no way we know to get rid of it, although there are methods to reduce it.

If you want more accuracy in your movements—a smaller signal, and thus a smaller amount of noise—the answer we all seem to tacitly know is to move more slowly. This speed-accuracy tradeoff, as first suggested by Robert S. Woodworth in 1899 and later computationally described by Paul Fitts in 1954 (when it was dubbed Fitts' law), is as firm a neuroscientific doctrine as the field has produced. You can sprint to the door, but in order to get your key into the keyhole to unlock it, you need to move with a little more intention and precision. Deliberateness is a natural counterpoint to the noise that afflicts the signal to every movement we make. And yet, it is not an entirely satisfying response. We all know people who are both fast *and* accurate in their skilled movements. They have practiced long and hard to improve their movements and attain a desired level of speed and accuracy. Their practice has enabled them to reduce the variability caused, in part, by noise.

And yet still, there may also be those who *begin* on a different level, who innately, for some genetic reason, carry within them more efficient, less noisy pathways of transmission, from the neural connections in their heads to the nervous signals coursing down their spines. These lucky few, Wolpert says, might be the ones we pay millions to watch perform.

As the son of an evolutionary biologist, Wolpert had an interest in knowing the limits of motor performance on a vast time scale. And so he often wondered why humans move the way we do. When we reach out our arm, there is an infinite number of possible trajectories that the arm can take toward its target. Yet humans typically move in a highly stereotyped manner. We all reach for a glass in roughly the same way. If I ask you to imagine how a young toddler tosses a ball, your image will just about be the same as mine: stiff upper torso, no rotation, pushing outward with the elbow. Is that because they are taught that way? Or is there a fundamentally *basic* way to move? Most of the motor control literature up until the late 1990s suggested that movements were determined based on energy cost—the jerkier the movement, the more energy inefficient it would be. But Wolpert didn't entirely buy that. The literature didn't explain why the central nervous system evolved to actually care about being smooth.

The general concept of noise in our motor system has roughly been around since Woodworth, then a PhD student at Columbia, showed that fast movements are less accurate than slow ones. "It is clear that the study of accurate movement must consider at every step the speed of the movement," Woodworth wrote. In his rudimentary experiments, which required participants to repeatedly draw a straight line, he found that as the number of movements per minute increased from 20 to 200, the number of errors increased sixfold. Equal increments of speed did not produce equal increments of error. Movements at 40 per minute were about as accurate as movements at 20 per minute. Movements at 140, 160, 180 and 200 per minute were all about the same in accuracy. At 140 on, the error rate would hardly change if the actor was blindfolded. Woodworth called it the "bad effect of speed." He surmised that accurate movement was not just about intention but also about adjustment—tiny corrections of mistakes, an impossibility if the movements are made at rapid speeds. "The path to skill," he wrote, "lies in increasing the accuracy of the initial adjustment, so that the later groping need be only within narrow limits." He gave the example of singers who managed to guide themselves by the sense of pitch or violinists finding their notes. Years of practice have enabled them to locate the grooves more quickly and easily, without the "groping" of a novice, Woodworth said. But this is just to mask the inevitability of some immutable error.

The fact that most movements were roughly the same but never exactly alike was captured by the Russian physiologist Nikolai Bernstein, who gave it a rather elegant description: "repetition without repetition." Outcomes may be replicated, but no two movements are exactly alike. When you move your hand quickly from one point to another (reaching for a coffee mug, for instance), it will follow a common trajectory, almost straight but slightly curved, velocity rising and falling in a parabolic distribution. Do it again, repeatedly. It should retain *almost* the same trajectory. Almost, but not quite exactly. Researchers knew this was not just because of the mechanical properties of our muscles and joints. Noise in the firing of motor neurons (the command signals to the muscle) corrupts. But if that were its only property, then errors could theoretically be minimized by moving as quickly as possible, reducing the opportunities for noise to interfere. Yet we know that is not the case—the faster we move, the more errors we tend to make. Moving very fast can even be counterproductive. Pondering this, Wolpert and Christopher Harris, a fellow researcher at University College London, proposed a unifying theory in *Nature* in August 1998: Noise must also accumulate. They wrote excitedly, "Movements can be described as trajectories that minimize post-movement variance in the presence of signal-dependent noise." Noise causes trajectories to veer, enough so that the final position of the movement can be altered. It affects it in an interesting way: The bigger the command, the more noise you have with it. "We thought of that as being a fundamental limit," Wolpert said. A limit on performance. Maybe movement paths aren't about smoothness or cost of energy. They are about reducing the bad consequences of what is called "signal-dependent noise." This noise inherently imposes a tradeoff between movement speed and accuracy—the same tradeoff that Woodworth mused about 99 years earlier. Speed, in this case, was a bigger command, and a bigger command meant more noise. The explanation for Fitts' law was no longer about a desire for smoothness. It was about reduction of noise. We live in a soup of noise, distorting not just our motor signals but almost everything around them as well.

Noise is present at all stages of sensorimotor control, from sensory processing through planning to the output of the motor system. It afflicts the very cells passing the signals to command movements or actions. It muddles the channels through which those signals pass. It fluctuates the opening and closing of the gates to those channels. It builds up in neuron-to-neuron interactions like an inflating balloon. The noisy signal that starts in the brain spreads into a noisy signal that ends in the muscle. It is an inescapable consequence of the scale, sophistication and complexity of the networks we rely on to generate behavior through movement, which is what Wolpert argues is the sole reason we have a brain to begin with. The result, all told, is variability—in how we perceive, make decisions, and act.

The unstructured nature of noise makes life especially hard for control. The reason we are not stumbling around like drunkards all the time is that we have developed methods of compensating for the disruptions. We learn through practice not to eliminate the noisy consequences of our actions, but in effect to work with them. Expert pistol shooters, for instance, learn to synchronize their trigger pulls with the involuntary tremors of their noisy bodies, while novices try in vain to immobilize themselves. Armrests aid surgeons seeking to stabilize their hands. Golfers turned toward long putters to anchor the club against their torsos on the green. Digital cameras incorporate anti-blur technology to mollify our subtle shakes. Our ability to compensate for noise is so great that we have adapted not to notice its existence. It's like a limp we have all learned to live with.

Internally, too, we might have ways to adjust to noise's most deleterious effects. A study from Oxford, in 2009, took 48 adults and trained half of them to juggle over the course of six weeks. At the end of the training, an assessment called "diffusion tensor imaging" (DTI) was used to examine architectural changes in the brain's white matter, the bundles of nerve fibers that reside predominantly below the cortical surface. For the first time in humans, the researchers saw training-related changes in white-matter structure, reflecting the possible influence of what is called "myelination." Myelin is the sheath that wraps around a nerve fiber so that impulses don't leak out. It looks like a puffy down jacket for nerves. It forms an insulating casing that, as with bumpers along a bowling lane, eases the pathway of signals to the synapse to be transmitted to the next neuron. This conduction mechanism enables neurons to fire faster, theoretically speeding the loops between sensory input and motor output. Faster transmission in certain loops might make up for noisy transmission in others.*

But Wolpert would note that it takes a ton of effort—a month and a half of intense juggling training for some meager changes in the fine architecture of our white matter—just to find a way to reduce some aspect of noise. Myelination does not miraculously transform everybody into Tiger Woods overnight, and it is expensive. Myelin takes up space. "You don't have infinite space in nerves," the neuroscientist David Linden told me. "So you wrap up only a few of them." Some nerves, like the one that transmits the throbbing pain sensation you feel after you stub your toe, transmit information at only a few miles per hour, which is why it can take several seconds for the feeling to arrive. The information is not traveling along a heavily myelinated nerve fiber. "As an engineer, you'd like to make

---

* Another study done with mice, published in *Science* in 2014, showed that by genetically blocking the mouse's ability to produce new myelin, researchers could prevent the animal from learning a new motor task.

them all fast," Linden said. "In reality, you can't build that with what you've got."

In Wolpert's view, the fundamental limit is still there, bogging us down at every step we take. "My gut feeling is that this noise is fixed in the person and you can't really reduce this magically," he says. Practice is just a long way around it.

To see noise for myself, I visited Dagmar Sternad's appropriately named Action Lab at Northeastern University. Some years ago, she created an experiment based off a game called table skittles, a leisure sport once favored in British pubs. It is like a miniaturized version of tetherball, with a marble-size orb dangling by a string from a meter-long pole. You grab the marble and fling it around the pole. The goal is to time the release of the marble so it forms the right arc to knock down a pin on the other side. You drink. There are nine pins in the pub game. Sternad's version is less convivial: One pin is represented by a two-dimensional marigold circle on a screen in front of you. The rig is an apparatus shaped like a lug wrench and designed like something you might find at an arm-wrestling competition. You rest your elbow on this manipulandum that allows you to move the wrist back and forth, in the manner of tossing a Frisbee. With your outstretched hand, you grab a lime-size wooden ball positioned at the socket end and use your index finger to tap a trigger button. That enables you to grab the virtual marble. By releasing the trigger, you release the marble, hopefully at the right time to form the elliptic arc around the post to trip the pin.

It is not easy to do. I tried, and missed on my first five tosses. Ten tosses. Twenty. Postdocs were laughing at me. Inebriated Brits in a pub somewhere were probably laughing at me. Finally, I gave

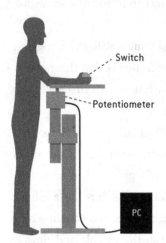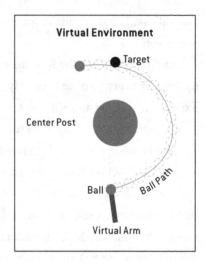

Simulation of a game called table skittles, popular in British pubs. Like tetherball, a ball is swung around a pole, but the goal is to knock down a single pin on the other side. Sternad uses the virtual game to study neuromotor noise.

up. "I can see now why you write about sports," Sternad said, helpfully.

My own struggles aside, the skittles paradigm is one that has yielded considerable insight into how we time our releases and how our own physiology conspires against us. Skittles is a nicely constrained task that contains only two variables that determine your "skill": the release position and the release velocity. It could in some ways be related to the timing necessary to intercept a fast-moving baseball pitch. Without noise, skittles would be easy. A few tosses to learn the path and a sense of the proper velocity, and we would be able to strike the pin repeatedly. Instead we are met with uncertainty. Fortunately, skill is not just about accuracy. It is also about redundancy. When Woodworth happened upon a blacksmith one day, he observed that the path of the hammer differed for each strike.

"And yet," Woodworth wrote, "the movement of swinging the large hammer—requiring as it does the concerted action of muscles of all limbs and of the trunk, each of which must contract in proper time and force—is executed with such precision that the hammer hits the drill every time." Somehow, he was coordinating the timing of the head within its downward blow. This is an extrinsic choice, implying a decision made by the actor about which path to take. Not that he is conscious of this process; he is just hammering away. However, his decision-making skill would seem no less poignant than that of the baseball players receiving neural assessments by deCervo.

What kind of decision-making is this? It could be argued that the blacksmith is deciding on the path that reduces the most noise, or what Sternad would argue is "most tolerant" of noise. Because total reduction is impossible, we seek ways to make noise matter less. In skittles, there is noise that impacts how we shape our movement to generate the optimal ball trajectory. The arm movement has a window of time in which we have to release the marble. Sternad has found that players begin by systematically exploring the range where small deviations in both the angle and timing of the release lead to only small increases in error. They test. They nibble. They find the window and try to live in it for as long as possible. They establish the tolerance of the procedure—how much one adjustment will get magnified in the outcome. Then they can adjust the shape of the arm movement as a whole. By exploiting this method, they can reduce the scattershot dispersion of their results, minimizing the impact of noise. She labeled this approach TNC (tolerance, noise, covariation). Players reduced the cost of T early on, followed by a slower decrease in the cost of C, ultimately promoting a modest decline of N.

The decline of N has riveted Sternad since she left Munich almost 30 years ago. She was then a popular radio host, television personality, and book author (*Gymnastik: Beweglichkeit, Kräftigung, Ausdauer für alle*, 1984) credited, at least partially, with introducing Germans to aerobics. She was introduced to aerobics by her thesis advisor, who, as with many others, was introduced to it by Jane Fonda. Extra petite, with emerald eyes and frizzly, candy-apple-red hair, she has retained both the enthusiasm and the exactness of her high-stepping days. When she noticed a student knitting during one of her weekly meetings, she grew annoyed but also motivated: She wanted to see, after decades of abstaining from the practice, if she herself could remember how to knit (she could). That, in turn, provoked an idea for another experiment. One of her postdocs, over the course of several years, would go on to pursue a hypothesis about the retention of skills. He found that, in some cases, even fine-grained kinematic features were faithfully retained after eight years without any practice. Once we get good at reducing noise in a particular movement, we can stay good at it.

Skills might not decay so easily, but it is also hard to improve them. N is there at every turn. There is little doubt in Sternad's mind that different people have different levels of noise in their system. "Every individual is different," she told me. "The chances that they have the same degree of noise is basically zero." But that is not to say that all noise is inherently destructive. Some believe that its presence affords exploration, in the way, perhaps, that losing your house key affords exploration of the exterior windows that might have been left slightly ajar. Others argue that N allows for flexibility. If I'm *so* good at a task, I won't be as inclined to try new methods. Wolpert does not dismiss these ideas entirely. "I believe there's a good tradeoff," he says. "You don't want too much, but you don't want too little. I think in general, though, noise in the system does affect our ability to do accurate control. If you have a robot with no noise, it can make precise movements over and over again. We don't do that."

Sternad is driven to understand how we might use noise and why our motor system is so fallible because of it. First, however, she feels it still needs to be better defined. Recently, she attempted an experiment motivated by microsurgery to determine how steady a person can be with an instrument. Magnification of the target, via microscope, helped the participants maintain some level of control. But the degrees of shakiness were staggering. "The individual differences were huge," Sternad said. Some of that, she said, was probably due to error correction: moving the tool to the target after magnification effectively illuminated the path. But her hunch is that these types of minute, conscious corrections (based on feedback) can be experimentally detached from unconscious noise, revealing the truer nature of N.

Are there some people who are blessed with an advantage over others based on the amount of noise in their motor system: brain, spinal cord, and limbs? "My guess," Wolpert says, "is probably." Though Wolpert was not thinking at all about athletes when he worked out his model for signal-dependent noise, it is not hard to see where noise might factor in the equation that determines who can go on to a lucrative athletic career and who is better fitted for a desk job. We all know people who just seem preternaturally gifted in a way that makes their movements appear smooth and tranquil. Their actions are both fast *and* accurate. They are the natural athletes, the ones who seem on a fast track to achievement in almost anything that involves a motor skill. I think of my best friend growing up, Dave, who could pick up a pool cue and, with almost no practice, run the table in a fashion I wished I could replicate. He made it look easy. And for him, maybe it was.

There are many factors that must come together for a performer to reach an elite status, and the notion of deliberate practice has been demonstrated to be as plausible a ground rule for achievement as any we have (although there will be more on that in the next chapter). The innate functioning of our neural circuitry does not prove that some performers are purely born and others purely made. But it should not come as a shock that our motor systems—like the brains and bodies they inhabit—are volatile, vulnerable, and far from uniform. If a motor act comes easily once, perhaps that is all the incentive one needs to keep trying it for 10,000 hours until it is perfected.

From Wolpert's office, the windows face east, spanning across a vibrant village of pubs and bookstores, cobblestone streets, the 213-foot Gothic Revival steeple of Our Lady and the English Martyrs, and a swath of green lawn called Parker's Piece, which serves as a communal park. There, in the early nineteenth century, students and townspeople used to gather to kick around a soccer ball. In 1848, they decided to write down a few key rules of the game. The first official rulebook of the Football Association was adopted from those notes in 1863—modern soccer, for all intents and purposes, was born. Today, soccer fans love to debate about who is the best player in the world, Lionel Messi or Cristiano Ronaldo. The popular narrative is that the two superstars arrived at their pedestals from differing paths. It is not fair to say that Ronaldo has worked harder for his success, and Messi has more natural talent. But that is sometimes the way it appears.

Wolpert likes to joke that he is the least computational person in the Cambridge engineering department. After getting a medical degree and earning his doctorate in physiology, he thought he might study population dynamics until his mentor, John Stein, told him, blankly, "Population dynamics is dead; neuroscience is the big thing now." He was researching motor control at University College London when the head of the medical school at Cambridge, Keith Peters, called and informed him they had an opening for a chair in engineering. "I think you've got the wrong person," Wolpert said. "I'm not an engineer." Peters responded, "We're not idiots at Cambridge. We know what you are."

They wanted to start a bioengineering department, and instead of hiring an engineer with an interest in biology, they settled on a biologist with an interest in engineering. When I visited, on a rare sunny morning in January, two large whiteboards in Wolpert's office were filled with equations. Dan McNamee, one of the postdocs, tried to assist in describing the focus of their research: "It's computational approaches to neural phenomena in control and perception." It was a lot to fit on a business card. Sufficiently confused—I expected neuroscience labs to be cluttered with assorted rigs, scanners, caged animals, fume hoods, and at least a whiff of formalin—I also happened to arrive on a Wednesday, which is one of the three days a week Wolpert likes to get everybody together in the kitchen area / conference room for a 4:00 P.M. "tea talk." It is a 10-minute informal presentation, given by a different person every time, about "something of interest to them but not something they're working on." It is meant to stimulate conversation and generate ideas between the biological researchers in the lab and the computational (or machine learning) researchers. I could not make out any distinction. On this day, the tea talk was presented by one of the machine learning graduate students, Alexander Matthews, who promptly asked, "How many people have heard about the determinant?" (a value used primarily in linear algebra). Only Wolpert raised his hand.

James Ingram, a senior research associate who has worked alongside Wolpert for more than a decade, explained the purpose of the work in the lab this way: "You can study the molecules, the cells, the synapses, all the gooey stuff. There are many levels you can study the brain. But if we have an engineering program, then we can build one. That's what computational neuroscience is about for me: having an understanding of the brain that gets us to build an artificial one." Jordan Taylor, a professor of cognitive psychology at Princeton, said Wolpert's writing style and algorithmic approach were transformative when his studies became hits in the late 1990s. "The engineering perspective in motor control was a tiny little niche," Taylor told me. "That little niche didn't have a lot of papers."

"And that's changed?" I asked.

"Today," he said, "all our stuff is written in that perspective."

Within the lab, Wolpert is known for two things: his cleverly analytical mind and the speed with which he himself moves. Specifically, the walking speed of his six-foot-three frame, behind which I struggled to keep up. "That's just him," said James Heald, one of the PhD candidates. "He's on a different pace." The only thing that matched his walking speed was his talking speed, although, at times, his tongue seems no match for his rapid thoughts and posh accent, so he clicks it between statements, sort of like a jump start. Bounding through the hallways of the lab, walking fast and talking faster, it is a wonder he has the patience to diligently tackle the dense mathematical computations he relies upon for his work. "I think he's primarily driven by his enjoyment of the whole process," Mc-Namee said. "I think that permeates the entire CBL," referring to the name of the lab, Computational and Biological Learning.

The ubiquity of noise in the brain and down through the neu-romuscular junctures to the limbs means that every movement we make will contain some degree of inaccuracy. Baseball hitters should rue it because it forces them to constantly calibrate the plane of their swing and relish it because it keeps pitchers from truly having pinpoint control. As our brain seeks to find a strategy for diminish-ing the consequences of the variability, the goal to keep in mind, Wolpert says, is that we ultimately want to make movements to achieve a rewarding state (or remove ourselves from a punishing one). The rewards we expect to receive and the costs we expect to pay should determine how quickly we move, what trajectory we decide to move in, and how the movement should respond to sen-sory feedback.

But, then again, the goal is not always the only thing in mind. Think again about how we walk. If optimization was characterized as a desire to never trip, we could be really, really careful with our foot placement upon every step. But that is not how we walk. When zoologists model walking, they describe it as minimizing energy, not minimizing injury. A trip-proof style of walking would be in-credibly inefficient. We don't mind tripping occasionally to save a lot of energy. In many ways, Wolpert says, human beings inherently "want to be lazy." Our brain strives to seek a balance between task success and not wasting energy. Some things require a certain amount of precision—bringing a fork to your mouth without constantly stabbing your lips is an example where laziness might be overridden by achievement (and avoidance of the cost of pain). But the brain is constantly making adjustments and approximations in order to limit how much energy it needs to expend. We have more than 200 joints in the human body. To flex or extend each of them would create 2^{200} configurations of movement, a number larger than the number of atoms in the universe. "There's no way any system can represent something so high-dimensional," Wolpert says. So you don't operate all 200 joints independently. The brain creates synergies. The hand has roughly 20 degrees of freedom, although in everyday use we only use about seven of them, which cover almost everything your hand needs. "That's why learning the piano is so hard," Wolpert says. "Because in order to learn the piano, your fingers have to work independently. We're used to doing things that are not so independent." He adds, "Evolutionarily, you want to use these synergies because they're useful to you, and also because they're easier to learn. But you can break them. With lots of learning, you can break them."

Most times, we do opt to just "go with the flow," as one researcher told me. Our brains *learn* to discover and exploit the passive dynamics of our systems, as well as the objects we manipulate. To illustrate this, Sternad tested an experiment on a man named Adam Winrich, who set a Guinness World Record in 2016 with 646 whipcracks with two whips in one minute. I personally cannot imagine going much faster than that, or why anyone would want to try. But in the Action Lab, she had him crack a three-meter whip against a target (a cat toy) spaced three meters away. By placing reflective markers along his shoulder, arm, wrist, elbow, hand, and three points along the whip, they could follow the path of his arm and the rope as it traveled from his side to the target. They noticed that this path was far more homogeneous and less variable when Winrich whipped the rope continuously, in a rhythmic pattern, than when he took a pause between each crack, a stop-and-start pattern called "discrete movement." Discrete movements are volatile. They are noisy. They reflect the brain trying to organize all the joints and muscles toward a concerted focus. But as he moved rhythmically, the whip-crack could almost be boiled down into one smooth task, transferring momentum across joints, as if his entire arm were a single hinge—or the whip itself. "What happens at the shoulder gets funneled all the way to the end of the whip," Sternad said. "It's an expression of smoother organization across the limbs." Winrich was much more accurate whipping the target rhythmically than discretely.

We exploit these stable and convergent trajectories to reduce our intrinsic noise. Pitchers wind up to get their bodies in rhythm. Basketball players dribble before their free throws. But given the uncertainty due to neuromotor noise, its unstructured disposition and its deleterious effects on accuracy, how is it that we are able to make stable and consistent movements at all?

Accurate movement relies on feedback, which reinforces the brain on its current state in the world, like flight instruments monitoring the airplane's status at every moment of its journey. But feedback is slow. It is much slower than you would think. Hermann von Helmholtz, like most scientists in the mid-nineteenth century, assumed that nerve transmission happened almost instantaneously, or maybe at the speed of light. But when he isolated a nerve and muscle in a frog's leg in 1850, he stimulated the nerve with an electric current, which provoked the muscle to contract. Eventually, Helmholtz detected that the stimulation-to-contraction lag was pronounced, such that he could even quantify it. He timed the transmission speed at 30 meters per second. He related the linkages to sending a telegram and waiting for a reply. "Just as the central station of the electric telegraph in the Post Office in Königstrasse is in communication with the outermost borders of the monarchy," Emil du Bois-Reymond echoed in a talk in 1851, "just so the soul in its office, the brain, endlessly receives dispatches from the outermost limits of its empire through its telegraph wires, the nerves, and sends out its orders in all directions to its civil servants, the muscles." The nerves were not the high-speed conduction pathways they had once thought.

We now know that nerve signal speed can reach 100 meters per second, or 223.69 miles per hour. Formula 1 cars have been known to attain faster speeds. The delay is also inflexible. It is why the International Association of Athletics Federations, which governs track events, considers a false start in the 100-meter dash to be any jump off the blocks that occurred within a tenth of a second *after* the firing of the gun. Since it is physically impossible to receive and respond to a stimulus in a shorter amount of time, any runner who breaks sooner could only have been cheating.* Now consider a game like baseball, where only about 100 milliseconds are required for the brain to take stock of the position of the incoming pitch and

---

* The reason track officials use sound, rather than a flash of light, to signal the start of a race is that the brain's auditory system processes faster than the visual system. But there will be more on that later.

produce a motor command to swing at it. "If feedback was instantaneous, then we could just use feedback," Ingram says. But it's not. So how could we possibly account for such delay?

A striking example of this dilemma took place in 2004 at an event called the Pepsi All-Star Softball Game, where top baseball sluggers like Albert Pujols and Mike Piazza stepped to the plate to try to hit pitches thrown by Jennie Finch, a star softball pitcher. "Try" was the operative word. Finch, throwing underhand, struck out Pujols on three pitches and downed Piazza and All-Star outfielder Brian Giles without so much as a foul tip.

The reason had little to do with velocity—Finch's pitches, delivered from a softball mound 43 feet away, equated to a fastball thrown 95 miles per hour—hard, but nothing extraordinary to a Major League hitter. Rather, it was the underhand angle from which Finch released her pitches that Pujols and the rest found so discomfiting. It was like nothing they had faced before. Their database of knowledge from years of practiced learning was useless. They had to rely on sensory transmission, and the delay was ruinous. They had lost their ability to predict the path of the ball.

So what we have is a woefully exploitable system, forced to make predictions, with slow signal speed that is susceptible to deviation thanks to N. We are built like jalopies for an IndyCar track. *It should seem a wonder that our brains enable us to ever hit the ball at all.* Judging what is around us, never mind what might be coming toward us, would seem like a function of almost pure conjecture. "Our minds should often change the idea of its sensation into that of its judgment, and make one serve only to excite the other," John Locke once said. But how to be so sure?

Wolpert found a way. Technically, it was the eighteenth-century philosopher Thomas Bayes who found a way. He just kept it to himself. An English cleric, Bayes was largely unheard of when he died in 1761. Only after his death, a friend discovered an intriguing mathematical paper among his possessions and sent it to an editor of the *Philosophical Transactions of the Royal Society of London*. The editor published it as it was: "An Essay towards solving a Problem in the Doctrine of Chances."

In it, Bayes puts a new spin on conditional probability—the probability that one thing will occur based on another thing that has already occurred. It is a spin that incorporates the element of bias, based on prior experience, knowledge or existing beliefs. It is a spin, in other words, that makes probability and prediction more human.

You don't know who was recently elected president, but you heard that there is going to be a major tax cut. Bayes can help you figure out which political party the president represents (without reading a newspaper). Or you open your e-mail to find a message from an unknown address saying you are owed one million dollars. Bayes helps you determine whether to think, "I'm rich!" versus, "I'm being scammed." Or let's say you are walking across a college quad and meet a guy named Rob, who seems shy and withdrawn. Would you be more likely to guess that Rob is getting his PhD in math or an MFA in theater? Bayes can give you an estimate about the future career of shy Rob, in case you just had to know.*

Shy Rob does have to know how to react to external events that might be happening more quickly than our own motor system can

---

* These examples are from videos on YouTube, where there are dozens of videos offering real-world explanations of Bayesian problems. Some are funnier than others.

reliably handle. In the 1950s, Bayes' theorem began gaining traction in computational laboratories for robotics and machine learning, and later behavioral psychology, and then as a method for understanding visual processes. Late in the 1990s, Wolpert and one of his postdocs at UCL, Konrad Kording, thought of a different application. They said, "Wouldn't it be interesting to see if it applies to motor tasks?"

Bayes' rule states that there are two sources of information: data (which is your sensory input) and prior knowledge (which is based on memories or beliefs). The prior knowledge factor recognizes that not all states are equally likely. They are instead represented by a probability between 0 and 1, where 0 is "I totally don't believe it" and 1 is "I believe it to be absolutely true." If I asked you if you believe that all male cardinals are red, your belief might be 0.95—pretty certain. There is also conditional belief, which is your belief in one statement given that you're told another statement is true. What's your belief that a die would show a six *given* that it has fallen on an even number? Since there are three even numbers on a die, my belief would be 1/3 (or 0.33).

The equation involves A (something we want to estimate about the state of the world) and B (the sensory input we get). The idea is to know the probability of different states of the world given sensory input, or P(A|B).

The equation to do this, as described by Bayes, is this:

$$P(state|sensory) = P(sensory|state)P(state) \div P(sensory)$$

P(state) is the probability of different states of the world independent of sensory feedback. For example, how probable is it that

I'll see a cat today instead of a polar bear? This is called a "prior belief." If your brain relied solely on prior belief, it would be like hallucinating all the time. So I have to ground my prior belief in reality. This is why it gets multiplied by P(sensory|state), which is the probability of my current sensory input given a possible state of the world. This is called a "likelihood." When we multiply these two things together we get what is called a "posterior." It's a way of combining prior knowledge with sensory evidence in an optimal way to generate beliefs about the world.

A better way to think about this is with tennis. When playing tennis, you have to decide where you think the ball is going to bounce in order to return it. Bayes' rule gives you two sources of information in order to make this prediction. One is sensory evidence: You see the ball coming toward you. But your senses are far from perfect. Furthermore, there could be spin on the ball, or wind affecting the ball flight, or it could be moving so rapidly that you don't get a clear idea of where it's headed. Instead of taking the sensory input as a certainty, you instead consider it a *likelihood*—the probability of seeing the ball bounce in a certain spot. Your brain uses a second source of information: *prior knowledge*. Tennis serves are not uniformly distributed around the court. They are likely to be confined to certain areas, areas that I know from experience are where my opponent likes to hit the ball. There's my prior belief. When I plug these calculations into Bayes' rule, I arrive at a different, predicted location of the ball. This may be how the brain navigates a complex and variable environment, sequestering the senses as only part of the equation. The system also relies on some prior belief based on a probability that what we have already experienced will be repeated. Athletes are reinforcing their prior beliefs whenever they watch film of an opponent before an upcoming match, memorizing and studying their tendencies so that what they see on the court will not catch them by surprise. These tendencies plug into a constantly updating equation for how to optimize their movement responses. Tennis players aren't just sensory-directed automatons: see ball, hit ball. They are statisticians running Bayesian equations in their heads.

That is how Wolpert and Kording's first paper on Bayesian Integration appeared on the cover of the January 2004 issue of the journal *Nature*, along with a picture of a female tennis player returning a serve. As she stretches toward the baseline, a splat of red is depicted in the area of the court where the ball looks like it could bounce, whereas a line of green represents a distribution of serves from the past. The target, a yellow circle, is where the player is headed with her racket. She did the calculation in an instant. "Scientists have found that the sort of movements and judgements performed by the world's No 1 tennis player closely match a mathematical principle first formulated in the 18th century," raved *The Independent*. It made sense to consider that the ability for athletes to make decisions under pressure—how to return a tennis serve, or make a football pass, or hit an incoming pitch—involved anticipation for how the scenario would unfold based on a strong understanding of probabilities. The estimate can be optimized if experience is combined with sensory evidence. I'll know where the ball is headed just as I'll know that an e-mail is probably not going to make me rich.

Now Bayes is everywhere in neuroscience. It is a model for answering questions about learning, perception, memory, reasoning, language and decision-making. The response of axons to molecular gradients has been described as "Bayesian." As Kording notes, many of these interpretations have strayed from what the researchers originally concluded. "On the one hand, the field took up that idea and countless papers have shown that behavior of humans is close to the one predicted by Bayesian statistics," Kording says. "On the other hand, the field interpreted it as the brain being built on Bayesian principles." Wolpert says it does not surprise him that aspects of the brain might be considered Bayesian in nature. "If you're not Bayesian, you're not optimal." A corollary view, the Dutch Book theorem, says that if you're acting in a non-Bayesian way, you will be exploited, which doesn't bode well for survival. "If you want to make the most accurate movements as possible, you've got to be Bayesian in some sense, otherwise you won't be efficient." Maybe it doesn't have to fit precisely into Bayes' rule, but it has to be close.

Wolpert says there is practically nothing you can't explain in a Bayesian way. How the brain actually does it is a more difficult question to answer. Like tennis, there are any number of illustrations of the way we make difficult calculations in our heads instinctually without needing an advanced degree in statistics. The economist Milton Friedman, speaking of how firms optimize their ability to do business, used the analogy that an expert billiards player is not explicitly calculating the angles on the table and the force on the cue ball. He just learned the proper methods through practice and experience, and thus behaves as if he did know the precise geometry. Likewise, Wolpert says, the brain is probably not making the strictly Bayesian calculations that fit into a simple formula. In order to properly construct such a model, you would have to integrate all the possible things that could happen and all the possible things you could have seen in a given state. "It's an equation of god knows how many dimensions," one researcher told me. It becomes intractable even for the fastest computers in the world. "We don't think the brain could do that either," Wolpert says. "It becomes, 'Can you efficiently do it and come up with approximate solutions that can be close to optimal?'" Not every tennis serve is judged correctly. The senses can be unreliable enough to throw off our Bayesian calculations entirely.

A perceptual phenomenon called the McGurk effect was first described in 1976 by Harry McGurk and John MacDonald, after asking a technician to dub audio of a woman saying the syllables "Ba Ba Ba" onto video of the same woman mouthing the syllables "Ga Ga Ga." When they played the tape, the subjects actually perceived "Da Da Da." The illusion is a multimodal one, tricking both your eyes and your ears. The reason you come up with a third syllable, Da, instead of either of the two correct ones is Bayesian. Bayes' rule attempts to figure out a process to come up with the best possible *belief* given the conflicting sources of sensory information. It turns out that, in that case, you get Da Da Da. Is that terrible visual processing? No. It is actually optimal. It just happens to be wrong.*

The human brain does not get fooled often, though. A popular truism in baseball is that Major League hitters will catch up to a fastball over the plate at any speed. In October 2016, an article in *The New York Times* interviewed hitters about what it felt like to face then Chicago Cubs pitcher Aroldis Chapman, the hardest thrower in Major League history. With a left arm that threw a pitch recorded at 105.1 miles per hour, Chapman seemed to defy the limits

---

* A 2004 study by researchers in Italy suggested that the same McGurk effect explains the trick exploited by ventriloquists. Our sensory apparatus is biased toward vision, which is why we fall for the puppet's moving lips rather than the source of the actual speech.

of physical capability. But, as the *Times* noted, of the 26 pitches he threw at 104 miles per hour or more, 12 resulted in contact and only one resulted in a swinging strike. "You can't even find a pitching machine that goes up that high," Baltimore Orioles infielder Ryan Flaherty told the paper. "You just start early, get going, try to pick up the ball. It's milliseconds." And still, Flaherty was able to make contact with a 104.9-miles-per-hour pitch and ground out.

His brain found the solution. In a 2011 *Neuron* paper, Wolpert and David W. Franklin described the computational mechanisms the brain of a goalie might use to stop an oncoming hockey puck. There was Bayesian theory to help handle the uncertainty of the location of the flying puck. A model to describe how he found the optimal set of muscles to move into the right positions. A predictive mechanism to compensate for the delays in the sensory and processing system.

And, perhaps, this was a goalie blessed with a little extra in his favor: neural circuitry that enabled the messages to the muscles to be transmitted a little cleaner, a bit less noisily. I would pay to see that.

3.

THE MOTOR HUNTER

WHY STEPHEN CURRY IS A GENIUS

O ne July morning at the Santa Fe Institute, a flat compound of cornhusk-yellow buildings, nestled in cholla and slouched atop a rocky bluff in the foothills of the Sangre de Cristo Mountains in northern New Mexico, a group began to gather. Above, an unkind sun already glistened. Below, a vivid tableau of mesquite desert, and beyond that a sprawling metropolis of parking lots and big-box retailers and low mesa homes, and beyond that a wall of violet mountains surging up toward the sky: the Manzano, Ortiz, and Sandia, and the Chicoma volcano in the Jemez range, all dim and hazy, like wisps of smoke. At dusk, the sun takes its time receding below these peaks. The compound is nearly 7,500 feet above sea level, perched in quiet repose. It is a retreat for visiting scholars to ponder the complexity of the universe. Hikers occasionally stumble upon it, perhaps thinking it a convent. They might then notice the equations written on the windowpanes. When the invited guests arrived, a soft warm breeze blew across a shaded veranda. The attendees included two neuroscientists, three sports scientists, a geneticist, a psychologist, an aeronautical engineer, a mechanical engineer, a theoretical physicist and a music theorist. It was a meeting of minds to discuss the future of bodies. A professional aerialist came, as well as a renowned dance choreographer, Spider Kedelsky, who studied with Jack Cole. In the entryway, there were bagels, muffins and an ice bucket of Red Bull. An army special operations member admitted he was a little timid about public speaking. A trainer with the San Antonio Spurs took a seat between a professor from MIT and the chief technologist of a company that designs mechanical exoskeletons.

At the very least, everyone seemed linked by their curiosity. No one looked as though they had a firm grasp on what was about to unfold at a workshop with the cryptic title "Limits to Human Performance." At 9:15, David Krakauer, the president of the institute, welcomed the crowd, quickly noting the portrait of Isaac Newton by Alberto Escamilla above the fireplace in the back of the room. He then introduced Andy Walshe, Red Bull's "director of high performance." Barrel-chested and sun-kissed, an Australian with pale blue eyes, Walshe boomed out a few motivational words. "It's a very significant moment," he said. "The idea of moving the conversation forward on the optimization of human performance—this is really the first outreach we've done on this topic."

Since the Milesians in the sixth century BCE—and very likely before that—serious minds have pondered the physics of the environment, and the human body, and how fast and far one can go within both. There is, theoretically, a limit to human performance, although the goalposts seem to keep moving backward. The winner of the 100-meter dash in the 1904 Summer Games, Archie Hahn, ran it in 11 seconds; today that is slower than the record set among 14-year-olds. The four-minute mile was considered unbreakable until it was reached; now it is the two-hour marathon that sports scientists are confident can soon be conquered. But in this room, there was a blend of both excitement and panic that the march of progress was inexorably slowing down.

The name of the summit in Santa Fe was a double entendre. It was as much an attempt to understand what *is* limiting human performance as what the theoretical limits might be—to understand the other factors below the surface, whether in training, conditioning, psychology, sleep, technology, nutrition or neural signals. There seemed to be a general belief hovering in the room that, by gathering these myriad intellectuals together, by the sheer osmosis of their collective thoughts, some advancement would be made in the name of "sports science." So when the first speaker rose and walked to the front of the room for a leadoff presentation, the place tittered a bit. There was palpable excitement. Anders Ericsson, the dean of human performance science, is the man responsible for the widely accepted doctrine of deliberate practice, popularly known as the "10,000 hours rule." Better than almost anyone, he had managed to apply his brand of rigorous laboratory research toward a fundamental understanding of how expert skills are forged in domains as incongruous as radiology, violin, and the spelling bee. A tall, leonine man in his 60s, with a bushy silver beard, appearing somewhat ruffled, he was the only presenter in a jacket and tie. He spoke slowly and carefully, referring occasionally to slides on a projector screen, which enumerated his fundamental keys to achieving an elite level of performance:

- Activate your genes through practice
- Employ teacher-guided deliberate practice
- Develop mental representations in order to hardwire those skills in the brain

After 20 minutes, he concluded to rousing applause. It was an engaging conversation-starter, rattling the minds in the room a bit while the Red Bull sank in. Promptly, questions sprang from Kedelsky, who wanted to know about the influence of a privileged upbringing on genetic makeup, and Andrew Herr, a professor of security studies at Georgetown. Ericsson's time was just about up. Another presenter got up to write his name on the whiteboard, indicating he wanted to speak next.

Finally, a hand shot up from the back of the room. The questioner identified himself only as "John, um, David's brother," which produced a few chuckles. His accent was British, although he happened to be born in Illinois and raised near Lisbon. He wore a tight black T-shirt and gray slacks and sipped from a Dunkin' Donuts iced latte. Throughout Ericsson's presentation, I had noticed, John had seemed only mildly engaged; occasionally, his focus seemed to drift, or he glanced at his phone, or he jabbed an elbow at the person seated next to him and whispered something in his ear. The night before, at a cocktail reception on the flagstone patio at David's home, John had said he intended to keep quiet during the workshop and had no plans to present. At this, David, knowing his younger brother too well, rolled his eyes. Sure enough, the first presentation was not over yet before John had something he wanted clarified. He raised his arm high into the air and stared down at the table while awaiting Ericsson's notice.

"Yes?"

"I think we have to admit that representation is one of the most vexed-about issues." John rambled, looking not quite at Ericsson but at some faraway point a little more to the left as he spoke. "The notion of mental representation is a little odd to apply to muscle fibers and how high you can get your leg up in the air. Not that there isn't a mental state in runners, but to talk about mental representations seems a little odd to me."

It was the first of many challenges, clarifications, interjections and interrogations submitted by John W. Krakauer, who proceeded to dominate the workshop by way of brute and unflinching criticism. He parsed the proceeding presentations like a trial lawyer. After objecting to Ericsson's reliance on mental representations, he scoffed at flimsy interpretations of mental states; probed the inconsistencies of "perceived energy"; and lambasted blind faith in technology as a dangerous "fetishism" without nearly enough biological concepts to support it. He chortled at the Spurs trainer's pleas for pathways through the "data smog" clotting his e-mail inbox. He had a pithy metaphor or anecdote to relate to everything. He dropped references to Linus Pauling, the discoverer of sickle cell disease; Patient H.M.; umbrellas; squash; Candy Crush; Prince; and his work with NASA and Pixar.

In a room filled with certifiable geniuses, it was clear that many quickly grew annoyed by the outsize voice rising above the working minds trying to keep up. Eventually, some of them became amused. By the fifth presenter, one of the participants, an Emory heart surgeon, joked that he would "have to disagree with John, whose glass seems more than a little half-empty."

Everyone laughed. At lunch, Krakauer picked at a plate of strawberries with an affect of disillusionment. "Why are we all here?"

John Krakauer, director of the Brain, Learning, Animation, and Movement (BLAM) Lab at Johns Hopkins University, is not always so dissenting. In fact, he loves learning new things, especially if they can replace the old things he doubts. Spending time with him, as I've done on more than a half-dozen occasions, is a bit like hanging around a food critic: He can be irrepressibly nitpicky, but the meal should be entertaining. Krakauer smiles teasingly when he talks, with the edges of his lips curled downward in a sheepish manner and his eyes fastened shut. He can appear commensurately embarrassed about his intelligence as arrogant about it. In either case, he knows he has it, and so does anyone within earshot. At 50, he is bubbly and energetic, prone to speaking with his arms and reaching out to grab the wrist or shoulder of whomever he is speaking to. He wears tight shirts that reveal a physique toned by a personal trainer and competitive squash. He favors spicy Manhattans. "It's better than a Baltimore," he likes to joke.

Krakauer is known in broader circles for his groundbreaking clinical work in stroke recovery, as the cofounder of the Kata Project, a neurology-technology hybrid model built around an interactive video game that, in its early stages, has produced encouraging results for motor rehabilitation ("kata" is a Japanese term related to form and mastery). But in narrower neuroscientific circles, I have come to realize that his influence is perhaps even more critical. Krakauer considers his stroke research to be his mission, but like any mission, it rests its foundation in a question: Why does it take so long to learn to become good at something? If Anders Ericsson is right—and 30 years of research on the subject of expertise would suggest that he knows what he is talking about—why should it take as much as 10,000 hours to become an elite violist, or a professional golfer, or a chess grand master? Why can't the fundamentals of a skill be translated more quickly into expertise? Why can a motor task in the laboratory take only a few hours or days to learn and become proficient at, but expertise in the real world is so distant and difficult to attain?

The body and its musculature are complex, and inundated by N, and so it should take time to learn a skill, you might say. To which Krakauer would retort: "You need to practice to get better at French. That's not a motor task. But French takes a long time. And so does chess, and math, and violin. Once you start talking about things that take a long time, the divide isn't between the motor and the nonmotor. It's the time."

Krakauer has given this a lot of thought. An avid tennis and squash player, he has composed his movement laboratory around an engine of ideas, including how to rehabilitate stroke sufferers. But concomitant to his altruistic ambitions (medical rehabilitation) is a desire to illuminate the dim understanding of the complex movements made by the unimpaired. He sometimes calls himself a "motor hunter"—a reference to the Paul de Kruif book *Microbe Hunters*, but in the neuroscience domain. Today, he says, there is far more hunting than finding. There is a lot about modern science that vexes John Krakauer, not the least of which has been its snobbish attitude toward those who are able to move their bodies in extraordinary ways: namely, great athletes. For this, he blames our inability to properly consider issues raised in the philosophy of ancient Greece. But more recently, he blames Henry Molaison.

Molaison was the patient with severe epilepsy known to the

public before his death simply as Patient H.M. In 1953, H.M. received a bilateral medial temporal lobectomy. Afterward, he developed anterograde amnesia. He could not commit short-term memories into long-term memories. Often, he would forget events 30 seconds after they occurred. In 1962, psychologist Brenda Milner had H.M. perform a task that involved tracing the outline of a star with a pencil, with vision of his arm obscured, using a mirror to guide him. Each day he would return to the task, forgetting he had ever done it before. He would have to be reinstructed each time. Over the course of three days, however, H.M. somehow improved his ability to draw the star. H.M. apparently did not need the recall of short-term memories to learn and retain new motor skills. He did not need his hippocampus. His brain had already encoded the necessary messages to deliver to the muscles, just as it had encoded memories such as Pearl Harbor and the 1929 stock market crash.

From this, a hypothesis emerged about the sort of procedural know-how required to achieve a motoric task, be it tracing on a sheet of paper, riding a bike or hitting a tennis ball. Because of the notoriety of H.M.'s case, this hypothesis gained significant traction. It fit a classical view from Socrates, who distinguished between a "habitude" and an "art" in Plato's *Gorgias* dialogue. A true practitioner of art, he wrote, can achieve results only by using "knowledge" of all the intermediate steps. An expert should even be able to explain the role of each of these principles. H.M. was improving at his task—on his way toward becoming a "true practitioner"—yet he had no knowledge of what he was doing to achieve it. His skills, therefore, must be more emblematic of "habitude," something imprinted deep in the brain through repetition and conditioning that could be subconsciously retrieved, like out of habit. H.M. could draw the same way a pianist could hop behind a keyboard and play, motor memories guiding her fingertips. There was no contribution of cognition, because it was not necessary. A great athlete often struggles to articulately describe what it was that enabled him to make the amazing play that no one else deemed possible. And thus did motor expertise become regrettably divorced from thought.

When I spoke to Krakauer about this, one day in his office, our conversation drifted toward Stephen Curry, the star guard for the Golden State Warriors. His mind wanders in Curry's direction quite often. The two-time NBA Most Valuable Player fascinates him to no end, because he cannot explain him. The success of LeBron James, at six feet, eight inches tall and weighing 250 pounds of sheer muscle, sure, that can be understood. There are genetic gifts in that package that augur nicely for professional basketball: height, speed, strength, quickness. And yet, in Curry, the most dazzling basketball player in the world is no more than six feet, three inches and 190 pounds. You would not think twice if he arrived at your doorstep and delivered your mail. In fact, Curry, the son of a former NBA player, was considered to be too slow-footed and *unathletic* by scouts, so many teams passed on him in the NBA draft.

And so Curry seems to symbolize all that is unfair in the mischaracterization of jocks as purely the physical (and not mental) amalgamations of genetic fortune.

"What do you say he's good at?" Krakauer said. "What do you actually say he's good at? I mean, you know what he's good at. But why? Does he just have better joints and muscles? Does he eat better breakfasts? What do you actually give the credit for? If you were to test his strength, his speed, his reaction times, he would be no better than anybody else in the league. None of those variables work. If you had to come up with a set of measurements that you could give to people that have never seen him play, and you said, 'Here are measurements, pick which one is him,' you'd never come up with a measurement that would capture it. So what is it? No one knows. They can see it, because he's amazing. But what do you assign the credit for? Is it because he's very smart? They'd probably say no. So you choose what to put on the other side of 'very.' Very what? Very skillful? That's nonsense. That's as if to say he's very skilled at being skilled. So all you guys with all your knowledge about sport and you have no clue what you're talking about."

Krakauer has no problem with Ericsson's theory of 10,000 hours; in fact, he almost completely buys into it. And Krakauer does not buy into much. There is a big problem with it, though. No one has any idea why it takes so long, because no one knows what it actually means to be skilled.

Sometime around 2015, an engineer and YouTube personality named Destin Sandlin was given a bike that had been welded such that any turn of the handlebar moved the wheel in the *opposite* direction. It was meant as a joke. Sandlin, who said he had been riding bikes since age six, thought he could conquer it easily. Spoiler: He could not. When he tried, he looked like a small kid again, wobbling violently for a meter or so before spilling over. Sandlin grew frustrated. He took the bike home, and for five minutes every day, he practiced in his driveway. He expected he could learn it in a few days. It took him eight months. Then later, when he tried to ride a normal bike again, he could not do that either. The learning curve was much shorter—about 20 embarrassing minutes—but Sandlin still needed to concentrate hard on the task in order to accomplish it.

If it was so difficult to *unlearn* a motor skill, what should that tell you about learning it? For one thing, the complexity of riding a bicycle is often vastly underestimated. This is probably because almost all children ride bikes. Monkeys can ride bikes. The Moscow circus formerly employed bears that could ride bikes. Most people who want to learn to ride a bike can eventually learn to ride a bike. And then we never forget how. The skill can always boomerang back to us, as if we had evolved to do it, as if it was in our DNA.

Of course, the ability to ride a bike is not something we inherited. We've just pretty much got the learning of it down to a science. When you are just beginning, you carefully follow instructions. You concentrate hard on every movement, focusing on how to maintain balance while coordinating the churning of your legs. Eventually, as you get better at it, you can phase out those explicit strategies for a more automatized one. The instructions are called "declarative knowledge," while automatization is said to be "procedural." When your bike-riding gets to be procedural, you don't have to think about it; you just do it. The skill sinks in. It becomes, well, like riding a bike. Motor researchers have a phrase for this stepwise pattern of learning, which they call "scaffolding." The early instructions support the construction of the skill until it is strong enough to support itself on its own. "The importance of repetition until automaticity cannot be overstated," John Wooden once said, and coaches have adhered to this dogma ever since. Indeed, automaticity can be great: Movements might come more quickly and easily, you are less fallible to distraction, and your brain can perhaps be more efficiently used to solve other problems.

But there can also be a cost to automatization. That cost is the thing we call habit. In Krakauer's view, the reason Sandlin had such a difficult time with his experiment is not because he was following the wrong instructions or had never acquired the skill to ride a bike to begin with. It's because he had also acquired a habit.

Habit is "the price you pay" for becoming automatic, Krakauer's laboratory codirector, Adrian Haith, explained to me. They have found that habit is binary; once a motor activity gets to be habitual, additional practice doesn't make it more habitual. There is no gradation. You either fall into the habit, or you don't. So what is the problem? The problem, Haith says, is that habit is "an automatic retrieval of an action, even when you don't really want it." Even when it is wrong. Obviously, this can have a deleterious effect on skill. In Sandlin's case, as soon as the handlebars were reversed, the habit he relied upon to ride a regular bike with ease became the wrong one for riding a bike with reversed handlebars. To adjust, not only did he need to relearn how to ride the new bike (by following a new set of declarative instructions). He also had to contend with the intrusion of a habit.

Habit, that enduring concept that seized the attention of Aristotle, Cicero, Descartes, Spinoza, Hume. Augustine feared habits; Kant called them "ridiculous." (That he was known as the "Königsberg clock" due to the exactness of his own daily routines was beside the point.) William James grappled with the philosophy of habit, arguing that it "diminishes the conscious attention with which our actions are performed." This, he wrote, allows our actions to arrive more quickly. "The marksman sees the bird, and, before he knows it, he has aimed and shot. A gleam in his adversary's eye, a momentary pressure from his rapier, and the fencer finds that he has instantly made the right parry and return. A glance at the musical hieroglyphics, and the pianist's fingers have rippled through a cataract of notes." But James, too, remarked about a cost. "Not only is it the right thing at the right time that we thus involuntarily do," he wrote, "but the wrong thing also, if it be an habitual thing." This cost, Krakauer argues, is not typically considered in investigations of motor learning or skill acquisition. The most famous model of the skill-learning curve, proposed by Paul Fitts and Michael Posner in 1967, involved three stages: (1) cognitive, or the declarative stage; (2) associative, or the exploratory stage; and (3) autonomous, which is self-explanatory. A fourth, offshoot stage might be habit, which could cast any frustrated performer, of any skill level, back to the drawing board. Where coaching can often miss the mark involves the need to proceduralize the *right* things for the right tasks. If baseball players spend their time hitting straight fastballs in batting practice, would the first sinker they face in a game not induce the sinking feeling of the reversion of handlebars on a bike?

Additionally, there could be implications for studying the neural substrates of motor learning. When researchers talk about "action controllers," "automatization" or even "muscle memory," they are really referring to representations of an action or sequence of actions that get formed, reorganized and consolidated in our long-term memory. All movements are derived from a myriad of smaller motions that get strung together, like beads in a necklace. It is believed that the more I reproduce a movement, the stronger those selected neural activations become, the better defined my brain's representation of what the completed movement should look like. This informs my plan for the movement upon the request it be made again. The mystery of how and where these sequences might get stored is the focus of considerable debate, and another chapter. But that such representations serve as the "organizers of activity," as the linguist and computer scientist Luc Steels has put it, is widely viewed as a given.

Yet there remains a very critical question when it comes to actually learning a skill like riding a bicycle, tying your shoes, swinging a golf club or driving your car—all universally absorbable skills of varying complexity that, after practice, can seep into the tranquil domain of "muscle memory." Are we to suppose that each of these representations is composed uniformly, hoarded into a homogeneous locker within our memory, as a monolith? Whereby you learn the requisite movements, consolidate them and, when called upon, can reproduce them as a skilled action? Are we then to suppose that our ability to drive a rental car is the same as our ability to drive a Formula 1 car with a steering wheel that looks like this?

Steering wheel of a Formula 1 car.

This is where Krakauer again tries to cleave apart the traditional synonymy of skill and habit. Let's say you are asked to answer a simple question in arithmetic. A straightforward question—what is 7 × 9?—can be solved a few ways. You could add nine to itself seven times. Or you could have rote memorized that 7 × 9 = 63. The latter method is taking advantage of the cached or automatized knowledge from weeks or years of flash cards and practice. You no longer need to follow the instructions; you just hop on the bike and do it. The answer arrives quickly to the tongue because your brain has consolidated those multiplication tables into an easily retrievable, yet inflexible, memory. This is a cognitive memory just as you might have a muscle memory. Krakauer remarked to me once that he has forgotten the numerals for his six-digit ATM code. This doesn't leave him destitute—whenever he needs cash, he can simply go to the machine and his fingers will immediately, automatically, punch the right formation on the keypad.

This sort of rapid, seemingly reflexive retrieval system works fine for some calculations. Krakauer and Haith, though, would contend that a mathematician is more likely to rely on both, procedural *and* declarative knowledge, in answering a long and intensive math problem, a math problem that would arguably result in observers remarking about her skill. They argue that a motor skill should be viewed the same way. The hallmark of any skill, they say, is being able to do the right thing, and quickly. But getting there is likely to require a more hierarchical representation, involving both forms of knowledge, braided both for quick retrieval and for the nimbleness that skill so often requires.

Actually, a good illustration of the interaction between procedural and declarative was H.M. Remember that with almost no short-term memory, he managed to improve his skill at mirror drawing. But did he do so without utilizing any declarative knowledge?

Krakauer points out that, actually, H.M. did receive explicit verbal instructions each time he performed the task, and he was able to use that knowledge. In other words, H.M. might not only have been retrieving procedural motor memories. His ability to draw was just as likely due to the instructions, despite his forgetting the next day that he ever received any. More recent studies involving similar amnestic patients have shown that they could not perform the motor tasks without instruction provided each day. When it came time to mirror-draw, H.M. appeared to be following both a declarative policy and a procedural one. As a result, his skill improved.

In this, most of us can relate. When we learn any new skill, a coach typically gives us guidelines for what to do. To properly shoot a basketball, for instance, you must hold your feet shoulder width apart, flex at the knees, position your elbow underneath the ball, stabilize it lightly with the opposite hand and so on. As you get to be good at it, you no longer require the explicit instructions, the scaffolding. But does your knowledge of those steps simply get paved over or disappear? Hardly. Stephen Curry could undoubtedly explain the proper steps required to shoot, even if he does not have to explicitly rely on those steps for his own shot. Furthermore, the automatization of the basics—like his shooting form—allows him to use declarative knowledge for more sophisticated demands. Krakauer argues that any expert is incorporating both automatized components and explicit know-how into their performance all the time. A skilled tennis player like Serena Williams possesses a massive repertoire of shots to choose from when faced with an oncoming serve. Serena does not require explicit instructions—say, her coach shouting from the sideline—every time to help her select which return to choose. She has automatized enough so that this repertoire comes quickly and easily to her. This actually liberates her to *add on* the declarative bits that enable her to excel: strategies about her opponent, consideration for weather conditions, or additional focus on specific details like foot placement or wrist angles, the way you might plan the coming workday as you are shaving in the morning. Dynamic activities, like tennis, with unpredictable opponents and unpracticed scenarios materializing at rapid rates, would seem to require more than automatization. H.M. did. Even though he managed to mirror-draw the outlines, and his accuracy improved, if the daily instructions disappeared, he would not have remembered them, and he probably would not have acquired the skill. In Krakauer's view, skilled action "is action guided by ongoing accrual and improving application of knowledge of facts about an activity."

In his book *Peak*, Ericsson writes that the purpose of deliberate practice is to develop mental representations, and that, over time, the practice will make these representations "better." There is little to trifle with there. The efficacy of deliberate practice is that it should help you in getting from declarative to procedural. But you also need to automatize the right things. A dynamic task like tennis is not like riding a bike. There are a lot of aspects inherent to that skill that have to become proceduralized. Thus, there are a lot of aspects that can fall victim to the intrusion of habit. And this might be why it takes 10,000 hours to become an expert.

In several textbooks of cognitive neuroscience, motor "skill" is not even included in the table of contents. Thanks to H.M.—or, more precisely, thanks to those interpreting H.M.'s case—a gross mischaracterization of skill was accepted and perpetuated. This mischaracterization has deep cultural ramifications. "Sports are widely considered a motor activity," Krakauer wrote in 2013, "and are often contrasted with intellectual or theoretical activities. This divide is apparent in the notion of the 'dumb jock' and contentious debates about the role of sport on college campuses." In other words, we have been conditioned not to call LeBron James a "genius" in the same way as a linguist or a mathematician. In fact, many athletes are considered intellectually inferior *because* they don't possess the same aptitude in mathematics as they do in athletics, because one requires cognition and the other supposedly does not.

H.M.'s case only emboldened a view that existed long before he ever picked up a pencil. A certain philosophical bias against those working with their hands has existed since the ancient Greeks. After Plato, Aristotle drew a similar distinction between *episteme* (knowledge) and *techne* (skill). Up until the 1980s, a dominant theory of cognitive psychology was that mental processes were purely computational—the mind is an abstract, incorporeal, thinking machine, with no functional relationship between cognitive and motor systems. The metaphor served to imply motor action as simply an "aftermath" of cognition, unconscious and uninteresting. Krakauer has made it a mission to set the record straight.

As such, Krakauer has argued that motor skills are just as "cognitively challenging" as traditional brainteasers like crossword puzzles. He can see no reason why expert football players, who have mastered the complex skills necessary for success in their sport, could not also be successful at activities considered "more intellectual." Objections to this point could be traced to the Socratic argument that expertise requires the ability to explain to others how the skill is executed. Indeed, "some examples of knowledge do seem to be ones that characteristically manifest themselves in

94

verbalization," Krakauer writes. Others do not. From Wittgenstein's *Philosophical Investigations*:

> 78. Compare *knowing* and *saying*:
>
> how many feet high Mont Blanc is—
>
> how the word "game" is used—
>
> how a clarinet sounds.
>
> If you are surprised that one can know something and not be able to say it, you are perhaps thinking of a case like the first. Certainly not of one like the third.

"Maybe we automatically expect people who are geniuses as athletes to be geniuses also as speakers and writers," David Foster Wallace once wrote. "If it's just that we naively expect geniuses-in-motion to be also geniuses-in-reflection, then their failure to be that shouldn't really seem any crueler or more disillusioning than Kant's glass jaw or Eliot's inability to hit the curve." There is a well-known video of a three-year-old girl completing a Rubik's Cube in 70 seconds. It is an extraordinary feat. Would you thus have to say she is intelligent? A genius? The girl has yet to enter kindergarten. But few measures of intelligence are more universal than a Rubik's Cube. Perhaps, then, Krakauer argues, we should be evaluating intelligence simply based upon extraordinary things we can do with our brains. Becoming an all-time great professional basketball player would be one of those things.

Misconceptions, unfortunately, are rampant in neuroscience. You have probably heard somewhere that as humans we use only 10 percent of our brain's capacity. A recent Hollywood blockbuster, *Lucy*, starring Scarlett Johansson and Morgan Freeman, was built around this theory. But it is nonsense. We use 100 percent of our brain, all the time. There is no latent "reserve" lying fallow somewhere. This is how it has been since our brains evolved from the tinier vestiges housed within the skulls of our ancestral reptiles. Except, that is not true either. There is no fragmental trace of a "reptilian brain" within humans, lurking there in its primitive form. The basic structure of mammalian brains is at least as old as that of the brains of birds and reptiles. We just evolved along a different path. What that left us with is a brain that is extraordinary in relative size, proportionally larger than any other species', housing approximately 100 billion neurons. Amazing, right? Unfortunately, none of these things are true either. A few years ago, a Brazilian neuroscientist, Suzana Herculano-Houzel, tried to figure out who originally estimated that the human brain contained 100 billion neurons. Turned out, nobody knew. The figure—a round, easy-to-remember number, astounding without seeming implausible—simply ossified into fact over time. She figured out a way to verify it. She ground down the exhumed brains of animals using a "brain blender," a Tarantino-esque device that sounds just as macabre as it is. Herculano-Houzel devised a method to count the neurons in her homogenized brain soup. What did she find? The human brain contains only about 86 billion neurons. If you are thinking, *Well, the difference between 86 and 100 billion is basically negligible*, consider that the missing 14 billion neurons represent an entire baboon brain, with about 3 billion neurons to spare.

Herculano-Houzel began with a simple query—where did this 100 billion neurons number come from? She turned a lot of what we thought we knew about the brain on its head. Following her work from afar, Krakauer was impressed. "She asked a question that nobody wanted asked," he said. His questions are driven by observations of behavior, both human and animal, which he frets that neuroscience does not pay attention to closely enough. "Take an animal hunting. How is it that an animal that is hungry is willing to creep up really slowly on its prey? When it must be dying to just leap at it," he said. "How do you have a system that is both accurate and strategic? How can you be both slow and fast? Patient and impatient? Cognitive and motor?" He was captivated by the incongruities in such a powerful, efficient biological machine. "You watch those wildlife programs not even sure why you're fascinated by it. But it's this conglomeration of computations and abilities that are unbelievable that they've been combined in such a way. We want to know how."

So, assuming it does take 10,000 hours to become an expert performer, a question Krakauer then asks is: Who in their right mind would actually want to do that? It is *a lot* of training—years of *deliberate*, excruciating, mind-numbing practice. Most people have neither the desire nor the discipline to keep sharpening their skills for such duration. Ericsson even admits that most of us remain at a hobbyist level of performance because we would rather settle at a level of skill we find replicable and enjoyable than continue to pursue further self-improvement. What sets those others apart would seem to be some form of self-motivation. In behavioral psychology, motivation is a persuasive force. The desire to succeed keeps people pushing toward their goal, even if the gains are incremental and relative gains slow down as the time curve extends. Golfers can shave 10 or 15 strokes off their handicap in a year, but it takes much more time—and more intensive, focused techniques—to get from being a five handicap to being a scratch player. And yet, a famous study from 1959 reported the performance of an industrial cigar-rolling task. Workers in the study, mostly women, had produced in excess of 10 million cigars over seven years—a staggering amount. And they were still getting faster. Where does that motivation come from? Can it be quantified? And what role does it play in motor expertise?

Here is where Krakauer's mission and his motor hunting dovetail.

The Kata Project is Krakauer's vision for the future of stroke recovery. Its linchpin is a smirking, cackling, computer-generated dolphin he helped create, named Bandit. Splashing about in an indigo sea (on a flat-screen television in the BLAM Lab), he has shimmering, Nemoic eyes; a wide, ivory belly; and an unwavering, cetacean smile. You want to join him on an aquatic adventure. And this is crucial for the human players who are asked to maneuver him among tasty fish and confrontational sharks beneath the waves. The players are patients who recently suffered a debilitating stroke.

Bandit was also designed, meticulously, to mimic the precise kinematics of the underwater movement of an actual Atlantic bottlenose. His realism, and that of the other creatures floating within his world, is critical. In order to effectively move Bandit, the players, using a robotic sleeve that functions as an arm-length joystick, have to effectively move themselves, which is no easy task for a recent sufferer of a stroke. This struggle to forge a relationship with a realistic avatar could be the key to unlocking motoric recovery in a player's limp limb. Indeed, there is some evidence that Bandit helps patients rehabilitate faster and more effectively than traditional physical therapy, reestablishing the connections in the neural pathways of the motor system that had been devastated during trauma.

Krakauer, who was schooled in medicine, finished his residency in neurology at Columbia at the same time he was completing a research fellowship in motor control. There was something about traditional stroke rehab, in particular, that nagged at him. Krakauer began to question if physical therapy was doing enough for stroke victims, who number around 800,000 in the United States every year.

The conventional therapy routine begins slowly, escalating in intensity after several months post-trauma. Many approaches focus on coping rather than repairing, like helping a patient to learn to write with his left hand now that his right arm is paralyzed. Even then, the most acute workouts are relatively benign—three hours a day for a couple of weeks or so. Compared with the workout of a professional dancer attempting to learn a new routine, rehabilitation seemed, to Krakauer, mild and uninteresting. But why should they be so different?

Krakauer thought of stroke along the same spectrum as elite athleticism, only at the opposite end. The same regions of the brain needed to be exercised in order to reproduce equivalent motoric responses. He thought of it like a driver who has to learn to maneuver a car without brakes. One approach is to relearn how you maneuvered the car in the first place. And yet, in therapy, *relearning* is often treated differently than learning. Most patients don't want to struggle through the steps of walking again; they want to walk. And that's the problem. The brain needs the rigorous work and the repetitions, the type of effort a young basketball player puts in during the summer to learn how to effectively use his left hand. In Krakauer's opinion, most people don't appreciate the amount of time and effort athletes and performers invest in their craft, because they only see the outcome. And they don't think of it as the same thing as rehabilitation, because basketball is fun. And so here is Bandit.

Krakauer purposely landed on a cartoon dolphin in an interactive video game because he felt the player needs to have some sentimental connection to the task in order to want to practice long enough to actually achieve results. "I don't think any other animal, maybe other than horses, has such a universal appeal throughout human history," said Omar Ahmad, the director of the Kata Project. Kata is, in essence, a video game company in a neuroscience lab. Ahmad, along with Promit Roy and Kat McNally, designed every aspect of the game and its subsequent app, Bandit's Shark Showdown. "You go back to the ancient Greeks and they talk about dolphins playing with humans," Ahmad said. "There's an emotional context." Controlling him is not meant to be easy. "The overall picture of this project is to put you in a foreign motor space," said Ahmad. "The intention of your hand is now going to control an animal that has dynamics. You have to learn and explore and figure out how it moves and how it navigates. The goal is very rich, nontraditional movements for motor recovery after stroke."

Players interact with Bandit by placing their impaired arm into a sort of exoskeleton, which provides some robotic assistance and gives the arm six degrees of freedom—forward/back, up/down, left/right, yaw, pitch, and roll. This enables happy Bandit to swim liberally throughout the water and even do flips above the surf. What distinguishes Bandit's game from others, Ahmad said, is how developed and professional it feels. There are 120 levels. A soothing orchestral piano and windpipe soundtrack. Bubbles. Hammerhead sharks. A whip of Bandit's tail produces a crack of firework-like explosions. It should come as no surprise that developers from Pixar have visited the laboratory more than once to design plans for a future collaboration.

It is an altogether new approach to physical therapy: No two movements in the exoskeleton are exactly alike. Every move is arbitrary. There is very little assistance. But there is *motivation*—you want Bandit to gobble up those tasty fish. These aspects exercise not only the muscles, but also the brain pathways that once had control of them. "Traditional PT is like muscle-lifting—you're doing repetitive things to strengthen your muscle," said Roy, the game's co-creator. "But if you're going to strengthen your brain, it needs to be more like a challenging cognitive exercise." Krakauer relates it to childlike exploration—reteaching the brain how to learn to move, as if some internal clock had been scrambled and set back to age two. There is a reason he uses the term "motor babbling," which refers to the short, arbitrary bursts of motion that eventually build into a smooth action like grasping or walking. Children don't learn to move while they're alone and miserable in a hospital bed. They are challenged and rewarded, physically and cognitively. They delight in the exploration of movement. Stroke therapy could do the same thing. It could treat patients like babies.

Krakauer's research in motor control also gave him familiarity with neuroplasticity, how the brain can change even in adulthood. Many stroke patients, as we know, do get better. Over time, their brains self-repair; they rewire. What the data told him, though, is that the bulk of this reparation took place within the first month after the stroke, a time of particular sensitivity and vulnerability for the brain's impacted region, but also a fertile moment for regrowth. Krakauer began to think of this period of hyperplasticity as a "critical window" for stroke recovery. The problem was that many hospitals were limiting their patients' access to activity in that period. "We have this window," he has said, "and we're not using it." The hospital environment seemed to be reducing the one element that John felt patients needed to exploit in the brain's window for recovery: play.

Consider Bandit again. The reason Krakauer went with a friendly dolphin and thinks that will be a better approach than a Nautilus machine is because of the motivational element. Players grow familiar with Bandit and are motivated to get him to succeed. In other words, it needs to be like practicing a sport. Patients should not want to stop at a hobbyist's level of performance. They should be motivated to keep swimming, developing expertise in a task that requires constant engagement of muscles they once thought were lost. They should want to play.

Determining how much influence motivation can have on a stroke patient's recovery could go a long way toward understanding its influence on an athletic field. In 2007, Krakauer coauthored a paper for *The Journal of Neuroscience* titled "Why Don't We Move Faster?" People naturally move at remarkably uniform speeds when performing everyday tasks, such as reaching for a cup or lighting a match. But this is not just because of the speed-accuracy tradeoff; you can move faster and still not lose accuracy at many of these tasks. There is another factor. Krakauer examined patients with Parkinson's, a disease that reduces their stores of dopamine, a neurotransmitter known for its role in the frontal cortex as the modulator of cost and reward. Without it, Parkinson's patients exhibit bradykinesia: Their movements slow down, because they struggle to scale speed with distance. This might not have everything to do with accuracy, however. Parkinson's patients might not effectively "select" their movement speeds because they cannot effectively weigh their costs. Dopamine could thus serve as an arbiter of implicit values on movements, which manifest themselves as intensity, or "movement vigor"—speed, amplitude and frequency. Motivation, as the fuel of dopamine, is not an instigator to the brain just to work harder, but to work faster, too.

"If you have two athletes with the same level of practice and skill, on the day that one is more motivated, he will shift his speed-accuracy tradeoff more than the other one," Krakauer told me. It is a short-term modulator on the long-term practice effect. Players can practice all they want, but in some brief, ideal circumstances, motivation can give an extra boost. A negative modulator, such as nervousness, can have the same effect in reverse. "When you talk about having a natural ability," Krakauer said, "it may not just be a natural ability in the skill. It may be an ability to do better with your modulators." To modify the modulators, so as not to be too confident or too nervous, you might need to practice. It's reinforcing something different from a physical drill, but it might benefit your performance nonetheless.

"What is the home-court advantage?" Krakauer asked. "It's motivational. Are you more skilled at home than away?"

Of course not. Skill is more than just the speed of your legs or the breadth of your biceps. Its secret formula might not be inside a Gatorade bottle, but in the dopamine coursing through the substantia nigra to the neostriatum of your brain.

The son of a German Jewish immigrant who served in the U.S. Air Force, Krakauer was exposed to a lot of the world at a young age.

His mother, Wendy, is British, which is how he developed the accent. But he spent much of his upbringing along the southern coast of Portugal, in a small fishing village called Albufeira. He loved literature and Wittgenstein. He says he could have easily become a history professor. But he was also interested in the brain, and what ailed it. As a resident at Columbia, training in neuroimmunology, he once attended a guest lecture by Richard Frackowiak, a well-known stroke neurologist from London. Krakauer, not surprisingly, raised his hand to ask enough questions that Frackowiak began to grow weary. But Claude Ghez, a prominent name in motor control and physiology, took notice and invited Krakauer to visit his lab.

When Krakauer showed up, some time later, he was surprised to see researchers experimenting with tasks that seemed to make sense to him. This did not fit his conception of neuroscience. "I thought test tubes and molecules and such," he told me. Instead they appeared to be "quantifying movements." He wound up doing his fellowship with Ghez and remained in his lab for seven years. He quickly distinguished himself as a bright pupil. He still responded to the tug of immunology—in 1997, he wrote a review of Jerome Groopman's book about working with dying cancer patients that appeared in *The New Republic*—but more and more grew accustomed to his new life as a motor hunter. In 1999, Ghez asked him to coauthor two chapters in the fourth edition of the influential textbook *Principles of Neural Science*, edited by Eric Kandel, James Schwartz and Tom Jessell. He spent the summer at Ghez's home on Fire Island, reading up on the past and present of motor control theories. "I didn't know shit," he said. He tore through the seminal works of John Hughlings Jackson, Charles Scott Sherrington, Nikolai Bernstein and Hans Kuypers. Together, he and Ghez produced an effusive introduction to the motor system, with soaring prose uncharacteristic of a neuroscience text. "The accomplished pirouette of a ballet dancer, the powered backhand of a tennis player, the fingering technique of a pianist, and the coordinated eye movements of a reader all require a remarkable degree of motor skill that no robot approaches. Yet, once trained, the motor systems execute the motor programs for each of these skills with ease, for the most part automatically." That was unmistakably Krakauer. He used similarly vivid imagery in another essay, 16 years later, analyzing a report on British medalists in the Olympics that determined psychosocial factors for athletic success but offered little toward understanding the neuroscience of skill acquisition. "By analogy, failing to wash one's hands, stress, and sleep deprivation may increase one's chances of catching a cold but do not fundamentally inform as to how cold viruses cause upper respiratory symptoms."

His writing demonstrated a knack for synthesizing dense concepts. He was a relative latecomer to the profession, which would come to shape his enthusiasm and, later, his cynicism. Early on, he worked especially hard, often staying late in the lab designing new experiments. But he would also look at an experiment and immediately begin to wonder if the design was flawed, even if it had been tried in labs for years. Then he would fret: Were the results misinformed? He would think to himself, *Why am I alone in this lab at 10 in the evening worrying about this stuff?* "I was actually annoyed with myself," he said. "There were years I was wondering, 'Why am I doing this?'"

Working with Ghez was a humbling experience as well. "He was savage with me. Savage," Krakauer said. "Nothing was right. My experiments were wrong, my writing was wrong, my reading was wrong, my understanding was wrong. He spent years beating on me and beating on me. I got more attention from a major scientist than probably anyone ever gets." Pietro Mazzoni, who also worked in the lab, said that Ghez might have had trouble relating to researchers who had not received a PhD. He seemed inclined to believe "medical school is training to be a doctor, and that's not the same thing as training to be a scientist." As a result, Krakauer put in double time, "teaching himself," Mazzoni said, "as well as listening to Claude." His intelligence allowed him to keep his head above water. In 1999 and again in 2000, Krakauer was the lead author on two papers involving reaching tasks that put him on the map in motor control. "He became part of a world that only knew him as a scientist," not a doctor, Mazzoni said. "I was with him at meetings and he would be standing with a paper and people would say, 'Oh, you're John Krakauer?' They were expecting some older man, wise, sciency-looking. Here was this young guy wearing trendy clothes." The overnight recognition gave him confidence and establishment where once he thought he was only moonlighting.

Krakauer's unique outlook put him on the right track toward understanding what he says he knows now: Motor research had been adrift. He started growing increasingly bothered by the literature he was reading, although initially he could not explain what bothered him. He just had a sense that things were wrong. "It was awareness that what we were looking at when we looked at patients or sports or humans didn't really match what we were studying in the lab," Krakauer said. "It always bothered me that it didn't really map onto each other. It was a bit of a con we were telling ourselves."

One of Ghez's most influential lessons was that discovery often follows noticing something that doesn't quite fit. The kind of thing easily dismissed might be a wedge into something new. Krakauer coopted that philosophy. "He finds the places where things don't quite fit," Mazzoni said, "and he goes into the cracks to figure it out." He can at times—such as the Santa Fe conference—come across as acerbic, almost purposefully agitating. But Eric Krakauer, his younger half brother, does not believe that it reflects a distrust of other people or parties. "John is not afraid to say something he believes in, regardless of whether or not it goes against the grain," Eric said, adding: "If you're going to have an intellectual joust with my brother, he's going to bring out the best in you. If you're in any way able to keep up."

His reputation as the community firebrand has only become bolder in recent years. Sometimes the very titles of his papers can be a giveaway to the forthcoming opinions: "Are We Ready for a Natural History of Motor Learning?" (No). "Recent Insights into Perceptual and Motor Skill Learning" (There haven't been enough). "Neuroscience Needs Behavior: Correcting a Reductionist Bias" (You're welcome). This last one, an 11-page scolding for the marginalization of hypothesis-based behavioral experiments, in favor of technological shortcuts, appeared in *Neuron* in February 2017. I knew Krakauer was excited about it because he had e-mailed me an advance copy two months earlier, then later tweeted about its publication with the message: "Our provocation is out." It seemed he had been waiting months, if not years, to unload. The paper was widely covered, with write-ups in *New York* magazine and *The Atlantic*. Anyone who knew Krakauer also probably felt they had heard it all before. "If he reads something that is rigorous but has one crack in it, he won't sleep," Mazzoni said. "That crack bothers him."

Krakauer has told me that he thinks neuroscience requires "a mixture of hubris and humility—the perfect blend of wildness and discipline." In other words, you need the hubris to ask the difficult questions and the humility to understand that you might not get the answers you want in your lifetime. "When I talk to relatives at home who don't really know all that much of what I'm doing, they're sort of expecting I'm doing all sorts of crazy experiments and stuff, or they say, 'Oh, why don't you put them in the brain scanner?'" David Huberdeau, one of BLAM's PhD candidates, told me. "I think the answer to that question—and it's something that's taken a while to appreciate—is that the kind of work that John does and insists on pursuing is stuff that you can be quite definitive about."

One evening in the fall of 2016, I met Krakauer and Haith for dinner at the Four Seasons along the northeastern rim of Baltimore's Inner Harbor. It was a drizzly night, misty and raw, the kind of weather that might have felt inspirational to Edgar Allan Poe, a former Baltimore resident. Krakauer was coming from the gym and took an Uber. He selected a low table near a murmuring fireplace in a quiet section of the bar area, which was busy for a Tuesday. The fog shrouded any view of the water.

In between bites of poached monkfish and braised bok choy, Krakauer began to ruminate on the burden of proof that stretches outward like a ravine between him and certain fundamental truths. These are the "inevitable" principles, beyond the "distorting lenses of the current time in history," as he put it. A false step sends him careening down another existential plunge. A year is wasted; five, ten. Even the simplest experiment—matching an error in extent with an error in direction—can become riddled with doubts: How do you know the brain measures equivalently in millimeters in both parameters? How should you correct those errors? How can you be sure that the difference is not one you actually manufactured?

Krakauer can be difficult. He knows it. He has even apologized to me for it. But by the time another glass of pinot blanc—their driest—had arrived, I'd already begun to see that, beneath his occasionally solipsistic behavior, his withering pedantry, was evidence of a much quieter and deep-rooted desperation. It is what compels him to understand, *truly* understand, the objective truths about skill and the motor system. He might not care who he pisses off in order to satiate it.

I asked him what he thought about the work by Jason and Jordan. I thought Krakauer might be intrigued by the idea of plumbing the minds of pro baseball hitters. But he began to worry whether baseball, as a man-made confection, was the appropriate key to unlocking some of the brain's most treasured motoric secrets. "Are we defining real principles?" he said.

"I sit in a room and I hear something being discussed, and nine times out of ten I feel that the central thing isn't being addressed," he said. "I've had this for 20 years now. Why do people like this stuff and think it's good and I don't like it and I don't think it's good? Like the elephant in the room is never addressed. I feel that all the time."

Neuroscience was becoming "this thing we all do"—go to conferences, write papers, apply for grants. It was losing its intrepidity. All the finger tapping, the feeble reaches, the cartographic expeditions in the scanner—this was not behavior. John Hughlings Jackson divined enduring wisdom about the motor system without conducting a single experiment; I could sense John Krakauer's wheels spinning the same way, as he groped for the right track. "It's about being intellectual rather than academic," he said.

"I think science always becomes something else," Krakauer said. "It's become engineering, it becomes maps, it becomes statistics. It keeps being not itself. If you look at what science is, minus the math or the engineering, it's more like the humanities. That's the deep dark secret: Science is not its instruments. That was something I knew far better than most people."

Two hours later, we got up to leave. I half-jokingly asked what else was keeping him awake at night. "In New York, they asked me that once," Krakauer said. "I said, merengue."

Back at the lab, I was strapped into a leather chair facing what looked to be an architect's drafting table, with a smooth, glassy surface. Positioned about two feet above the table was a mirror angled downward, reflecting a projection onto the glass. My right wrist was in a splint with three small circular air slats connected by a tube to a nearby oxygen tank. When the air came on, I could move my wrist as I would a pusher on an air-hockey table. The splint immobilized my hand as my arm glided around the table. There was nothing to aim for until one of the postdocs, Aaron Wong, cued up a program called the Barrier Task. A small blue dot appeared, shielded by a three-walled "barrier," looking not unlike the bird's-eye view of a solitary bald-headed man in a cubicle.

The goal was to move a cursor as quickly as possible from that dot to another dot randomly positioned no more than a few inches away. That other dot was similarly barricaded. I was not supposed to touch any part of either cubicle as I made my move, and I had to complete the action before time ran out. The time on the clock was one second.

There was yet another wrinkle to the task. As soon as I began the movement, both dots and both walls instantly disappeared. I was left staring at a blank screen, trying to remember where the target was located and the safest route to reach it in the shortest amount of time. As soon as I reached that target, another defended dot randomly appeared somewhere, and then that disappeared after I began to move again. You have to move around the screen quickly, as if swatting at invisible flies. Every move needs to be deliberate due to the splint; there is no subtlety. At first, as I struggled to make the connections, I thought the test was for my memory. But as I played, it became easier to visualize where the dots were located. I began to understand that I was really being tested on my strategy.

After the first run, Wong threw in a second wrinkle. Now, for just a split second before I began my movement, a line appeared displaying a pathway between the two dots. As I began my move, the path became seared in my mind; I could practically hear the voice from Google Maps telling me how to move.

With the path provided, my reaction times went down considerably.

My scores went way up. Wong was not really paying attention to my score. He was checking my reaction time—the milliseconds between when I saw the dots appear and when I moved my cursor, as recorded by motion sensors on the splint. With the highlighted path, my reaction time went down considerably compared to when I did not have the guideline available. "We think that that difference is the ability to plan the shape of the movement," he said. "In order to generate any movement, there are three steps. The planning, which is where you decide what you want to do and how to do it. The execution, which is following through on that plan. And then there is feedback at the end, which helps you update the plan for the future. So planning is almost the entirety of your movement."

Edward Evarts, back in the 1960s, found that changes in neuronal activity began to occur some 100 milliseconds before the onset of the movement. Neurons begin to fire. The *intent* to move alters the firing pattern even further. Electroencephalogram recordings can pick up shifts in cortical potentials in the supplementary motor area nearly a full second before a movement is made. The start of this period is known as the Bereitschaftspotential, or BP, which stems from the German for "readiness potential." Volitional movements don't get made in a vacuum, which distinguishes them from reflexes, the reactionary and stereotypic muscular processes that occur in vertebrates regardless of whether their head is attached to the rest of the body. Once the brain is involved, though, like a neurotic mother, it cannot help but agonize over different scenarios. The extent of the movement is believed by some to be planned even before the first action potential is fired down the spinal cord. It requires an extraordinary amount of calculation; in fact, an entire motor and sensory prediction mechanism. But thinking about the plan will often disrupt the plan. A tennis player does not consciously consider which muscles to activate in order to return a backhand. Her BP exists in a realm seemingly beyond the reach of consciousness.

Actual motor reaction time, though, can vary. The degree is not big enough to separate great baseball hitters. But there were still Stanford linemen who got off the ball quicker than others. What's interesting is where that separation actually occurs—is it in the plan or the command, the muscles or the BP? For more than a century, researchers had generally believed that the reaction is the sum of all its intervening parts—namely, the processing. The brain needs time to perceive and prescribe a response, and once it is ready, the scale is tipped. The reaction occurs. There is a limit to how fast one can respond to a stimulus because, as Helmholtz had showed, signals are slow. Your reaction arrives as fast as humanly possible.

But recently, Krakauer, Haith, Wong and others have demonstrated that this may not be entirely true. Reaction times *could* be faster. In fact, participants in some recent studies were actually capable of moving 40 percent earlier than their times would suggest. If they were startled by a loud noise, their times could be faster. If they had been moving quickly in previous trials, their times could be faster. If they were especially motivated to move quickly, their times could be faster. If they were asked to correct a movement already in progress, their times could be faster. For some reason, though, initiation speed did not seem to be a requisite for the motor system. In fact, there was a latency period. The motor system is designed to generate an accurate movement within 130 milliseconds of presentation of a visual target, and yet we involuntarily add an extra 40 percent of inexplicable lag time. In fact, participants in the study *could not* time their movement with the point at which processing was complete. This did not have to do with the speed-accuracy tradeoff, Haith says. A movement at about 200 milliseconds will still be as accurate as one at 300. Something else holds people back. "They're inserting some slack, for some reason," as Haith put it. The researchers began to think about reaction time as two distinct components: the initiation of the movement and the preparation. Preparation cannot be rushed, but it can hit a ceiling. Initiation arrives leisurely. In between, there is an empty void, a delay that could serve any number of purposes. It might be to minimize errors, to play it safe, to fully comprehend the instructions of the task or to allow for a change of heart. Certain risk-seekers might feel more comfortable with less of that padding. Perhaps their reaction times are naturally lower.

Whether that can be demonstrably trained or enhanced, or whether your reaction time is mostly inborn, remains unclear, but their work on reaction times has, in Mazzoni's words, "stimulated a lot of new research." Said Jordan Taylor: "We always looked at it. We never thought about it." The decoupling of preparation from initiation has also reinvigorated investigations into what actually goes into the preparation phase, which is exactly what Wong was testing with the Barrier Task. Krakauer has identified six critical processes that occur before movement generation; three "what" and three "how":

What

1. What does my environment look like?
2. What is my target of interest?
3. What is my task?

How

1. How will the goal be attained?

2. How will the movement look?

3. How will I make that movement happen?

The bulk of the reaction time is hogged by the "what," perceptual decision-making processes that support the formation of a "motor goal." But when Wong tested me in the Barrier Task, he gave me the "what." He also gave me "how 1" and "how 2," by showing me a path to follow. All my brain needed to figure out was how to make the movement happen, which should not be so difficult. Certain patients with ideomotor apraxia, often caused by a lesion in the left parietal lobe, struggle with this task, just as they would struggle if you asked them to pick up a toothbrush and demonstrate how they might brush their teeth, even though they would have no problem making a simple reach. A reach eliminates the "what"; they don't need to plan that movement. The processing of their "how," on the other hand, appears unimpaired.

When put all together, the voluntary generation of any movement requires the transformation of sensation into action, a transformation that takes time, although perhaps not as much as we thought. Because we instinctively buffer, our actions are separated from our sensations by what could be described as intention. The stimulus does not trigger a movement. It triggers a decision to move. We are not, as Haith says, "slaves" to our inputs, but functions of a smart and flexible motor system, which summarizes Krakauer's opinion about all aspects of moving animals, from hunting cats to stroke patients to the slow-footed Stephen Curry.

On a warm evening in April 2017, Krakauer sat onstage inside the Rubin Museum of Art, in New York, for a conversation with Patrick Vieira, the French soccer player who helped win three Premier League titles for Arsenal and a World Cup for France. Vieira is now the coach of Major League Soccer's New York City FC. Krakauer wore a Huey Freeman T-shirt under a navy blazer and casually impaled the prejudices that no doubt inhabited the minds of most of those attending: about habit, motivation, and the performer's brain being secondary to his body.

"A billion people watched the World Cup final," he said. "A *billion*. If an alien came down from outer space, how would you explain why a billion people watched grown men kick a sphere around?"

He let the question linger for a bit before returning to offer an answer a little while later. "What is it that makes a sport a sport?" Krakauer said. "I think it is that it contains the requisite degree of complexity." It is complexity that draws us into museums and enthralls us with chess and sustains us on the golf course, even though the performance itself can often be harrowing and miserable. But we would not do it if it were simple. "When we look at a bit of architecture, or a piece of music, or we watch soccer, we are actually admiring its complexity."

Too often, this notion gets lost on the same people who love to watch. "We're very schizophrenic when it comes to sport," Krakauer said. "Because on the one hand, a billion people watched the World Cup. On the other hand, deep down, we think that it means more to be a genius in mathematics or music than to be able to use the term for an athlete. What we've argued is that the distinction is a false one. There is as much cognition in sport as in any other endeavor."

After the talk, Vieira took pictures and signed autographs while Krakauer stood offstage, planning his next move. As they left the museum together, Vieira pulled Krakauer aside and said quietly, "We should talk sometime."

4.

"FROM MIND TO MUSCLE"

HOW THE MOTOR CORTEX WAS FOUND

In the spring of 1870, Gustav Fritsch, an anatomist, and a young psychiatrist named Eduard Hitzig began a series of experiments on dogs. They were Privatdozents, assistant professors, and at the time, in Berlin, no laboratories would offer them space to experiment on warm-blooded animals. So the men had to improvise. They performed their studies at home, using Frau Hitzig's dressing table as an operating surface. Their methods were just as crude; they began with no anesthesia, and the animals yelped and whimpered as the men sliced into their dura mater and peeled back the membrane sleeve enveloping the brain like an orange peel. One dog bled to death. They removed as much as half the skull, exposing the frontal lobes of both hemispheres. On a table nearby, they kept a chain of 10 Daniell cells (batteries) connected to a commutator and a pair of electrodes made of narrow platinum wires insulated with cork. The current strength running through the electrodes was so low it barely produced a tactile sensation on the tongue. But when they applied the current to certain portions of the cerebrum, they induced the traumatized animals into strange behaviors. They made the dogs move.

The notion that the movement of the body is controlled by the brain was hardly new. Ancient Egyptian medical texts refer to head injuries and motor dysfunction. In *Phaedrus*, Plato defined the soul—including the brain and mind—by one's movement: "Any body that has an external source of motion is soulless." Hippocrates examined paralysis, seizures and other disturbances to movement. "Men ought to know that from nothing else but the brain come joys, delights, laughter and sports," he said. "In these ways, I am of the opinion that the brain exercises the greatest power in the man." But not everyone agreed. Aristotle took a more cardiocentric view in midcentury 300 BCE, noting that the heart pulsed and contained blood, while the brain remained relatively inert. Blood was essential to movement and sensation, and the brain is cool while the heart is warm. Warmth distinguishes a living body from a dead one. The brain, he reasoned, must only serve as a cooling mechanism for the all-important heart. This view persisted for centuries. "Tell me where is fancie bred, or in the heart, or in the head?" William Shakespeare wrote in *The Merchant of Venice*. Anguished lovers still mourn their broken hearts, and loving parents still bestow lessons from the heart. The word "record" takes its root from the Latin word for "heart," *cor*, because the heart was thought to play a role in memory. We still think of remembering something by heart.

But some Renaissance thinkers slowly began to come around to the original Hippocratic advocacy of the brain's omnipotence. They quietly circumvented the Catholic Church's obedience to the dogma of Roman idols like Galen of Pergamon, whose written treatises in the second century CE on the nerves and pneumata (animal spirits) had survived as gospel for 1,300 years. They crafted their own conclusions, based on their own observations. As the prohibition against autopsy was lifted to help stave off epidemics of the thirteenth century, Mondino de Luzzi produced one of the first popular guides to anatomy, *Anatomia Mundini*. Leonardo Da Vinci created a cast of the ventricles by injecting hot wax through a tube into the brain of an ox. In 1543, Andreas Vesalius, a Flemish anatomist, published 25 illustrations of the human brain after careful dissection. He sensed that the body derived its life from the soul, which was engendered in and by the brain. "While on the one hand it employs this spirit for the operations of the chief soul," he wrote of the brain, "on the other hand it is continually distributing it to the instruments of the senses and of movement by means of nerves."

Then René Descartes, inspired by the new hydraulic moving statues in the royal gardens of Saint-Germain, offered a more comprehensive theory for the role of the Galenic spirits that flowed through the muscles via tunnel-like nerves. Descartes focused on the conarium, or pineal gland, which he believed acted as a switchboard, directing spirits toward the muscles in need of activation. The body, like the garden statues, passively moved at the whim of the soul, residing in the brain. "One may compare the nerves of the machine I am describing with the pipes in the works of these fountains," he wrote in *Treatise of Man*, "its muscles and tendons with the various devices and springs which serve to set them in motion, its animal spirits with the water which drives them, the heart with the source of the water, and the cavities of the brain with the storage tanks." The nerves were not truly hollow but composed of a "marrow" of thin threads, which, when set into motion by an external object, constituted the basis for sensation. The threads pull against an area of the brain, opening up the pores through which the animal spirits might flow on to the muscles. Thus, a movement—even as quick as a reflex—is sparked. "If for example fire comes near the foot, the minute particles of this fire which as you know move with great velocity, have the power to set in motion the spot on the skin of the foot which they touch, and by this means pulling upon the delicate thread which is attached to the spot of the skin." Advancing the mechanical metaphor, it is not unlike "pulling at one end of a rope one makes to strike at the same instant a bell which hangs on the other end."

Descartes' depiction of the pain pathway in the nervous system from *Treatise of Man* (1664).

Descartes was a philosopher, not a physiologist. The pineal gland, named after its resemblance to a pinecone, is a member of the endocrine system, not an incorporeal one. But Descartes helped usher in a new era concerning the brain. Of the parts of the body, "none is presumed to be easier or better known than the Brain," wrote a contemporary physician, Thomas Willis, "yet in the meantime, there is none less or more imperfectly understood." Willis, who coined the term *neurologie*, meaning "doctrine of the nerves," thought of the brain as containing various levels of functions. The more advanced functions—like memory for words and ideas—derived from the brain's uppermost structures, such as the cortex, where they could be stored in the grooves, or sulci, of the cerebrum. Dogs and fish, he noted, did not have those grooves. He grew fascinated by the corpus striatum, in the midbrain, which contained streaks of gray and white matter. Autopsies of patients with paralysis often revealed deterioration of the corpus striatum, "discoloured like filth and dirt, and many chamferings obliterated," as Willis remarked. Newborn dogs that struggled to move their limbs did not contain the same streaks as he observed in the corpus striatum of other species. Therefore, he presumed, that region of the brain must play a role in movement.

As Willis tried to functionally organize the brain, others tried to figure out the nerves, those narrow ducts supposedly whisking animal spirits about our mechanized limbs. Giovanni Alfonso Borelli saw the spirits not as wind or air, but as a juice, or *succus nerveus*, with "a liquid consistency like spirits of wine" carried throughout the nerves as if by a canal. The liquid, once deposited onto the muscles, produced a contraction. But in 1674, Antoni van Leeuwenhoek, a pioneering microscopist, made a critical discovery. When he examined the optic nerve of a cow (Galen had once remarked that he could see its hollow opening simply by holding the nerve up to sunlight), Van Leeuwenhoek could find no such mouth. The nerve did not appear to be hollow at all. Almost a century later, Albrecht von Haller attempted to tie up a group of nerve bundles to see if spirits, or juices, would swell up behind the knot. Not so. When he then cut the nerves, nothing trickled out. "Of what nature then is the material of these spirits?" Haller asked. "An element of its own kind unlike everything else." He called the element "vis nervosa" but was unwilling to go further in describing it. Isaac Newton attributed color perception to different particles of light causing vibration in the nerves from the eye to the brain. Nerves, he offered, might then transmit information by vibration. But he, too, was incorrect. The riddle of the brain's communion with the periphery left even Newton stumped.

Why was this secret so hard to decipher? Inquisitors had to venture far outside the mammalian class for a better clue. In the early 1770s, a scientist and member of the British Parliament, John Walsh, decided he wanted to investigate the existence of a particular species of fish, one that had become the source of legend: the electric eel. Walsh traveled to La Rochelle, on the Atlantic coast of France, and interrogated local fishermen who had had the misfortune of crossing its path. He later sacrificed his own arm to an eel's shock that traveled "half way of the part of my arm above the elbow; both instantaneous in commencement, and ending precisely as an Electric shock." In a note to a friend he expressed incredulity: "I exclaimed this certainly Electricity—but how?" He even discussed his puzzlement with another friend who had been studying electricity in America, Benjamin Franklin. One midsummer night, Walsh gathered guests at his home in London for another experiment. This time, Walsh (carefully) laid an eel on top of a tin plate, where it suddenly emitted a tiny blue flash of light. The guests were stunned. The news eventually made its way across the sea, where one politician regretted not being there to see it in person. On the day that Walsh made his discovery, his American friend had been signing the Declaration of Independence.

What Walsh had stumbled upon was a fundamentally new way to view the nervous system. If certain fishes could be "powered" by electricity, why couldn't human consciousness likewise get fueled by sparks in the brain? If nerves were not hollow pipes filled with animal spirits, what exactly were they filled with? In 1780, an Italian physicist named Luigi Galvani, almost certainly influenced by Walsh's discovery, began his own investigations into the existence of "animal electricity"—within the nerves and the muscles they traversed. With his wife, Lucia, serving as his assistant, Galvani in 1791 discovered that a frog's leg could be made to twitch when the nerve was touched by a metal scalpel held by a person near an electrical current, and, later, with simply a metal rod held between leg muscles and their nerves. The rod conducted the "electric fluid" just as the nerves should.

Galvani's *De viribus electricitatis in motu musculari commentarius* ("Commentary on the Effects of Electricity on Muscular Motion") redirected the course of physiologic exploration, revised the dictionary (we now have the word "galvanize") and also instigated some particularly macabre demonstrations. Galvani's nephew, Giovanni Aldini, carried out electrical experiments on decapitated heads he had collected from the guillotine, using impulses to produce different grimaces, move their jaws, and even open their eyes.

In 1817, German scientist Karl August Weinhold claimed to have scooped the gray matter of the skull of a kitten and replaced it with a compound of silver and zinc. For 20 minutes, he said, the animal managed to raise its head, open its eyes, crawl, hop around, and sink into a crouch. Weinhold was later found to be lying. But the influence of electrical stimulation—nicknamed galvanism—was enough that a young author, Mary Shelley, referred to the "physiological writers of Germany" in the preface to her 1818 novel, *Frankenstein*.

Almost 150 years after Hitzig and Fritsch began performing their grisly galvanic investigations on dogs in the living room, the electrical excitability of the cortex, and where and how certain regions might contribute to the production of movement, remains surprisingly relevant. At the Santa Fe conference with John Krakauer, I was surprised to see the name of an attendee representing a company I had been hearing more and more about in the world of sports. Halo Neuroscience, based in San Francisco, was founded by a Stanford neuroscientist and a Tulane biomedical engineer. Their product was a pair of sleek headphones, coolly fashioned to resemble the popular Beats by Dr. Dre. They were trying to get them on the heads of as many professional athletes as they could. According to Halo's website, the headphones were able to stimulate "the part of your brain responsible for muscle movement."

How this is done is with tiny nubs on the band that emit small electrical pulses targeted right for M1, the primary motor cortex. These pulses should interact with the electrical signals that already "galvanize" your entire central nervous system. Not to worry, it is a far less intense current than those used to trigger seizures in early electroconvulsive therapies for treating certain mental illnesses.

But the added boost from the headphones is said to kick those M1 neurons into some sort of productivity overdrive, aiding how quickly you learn new tasks and enhancing what you already do know by what the Halo folks call "neuropriming," or increasing plasticity in the brain prior to a workout. The engineer, Brett Wingeier, presented at the Santa Fe workshop and quickly noted his background designing something similar to implantable pacemakers in the brain that could reduce the frequency of seizures. Halo is all the same fuss without the mess. And not surprisingly, Krakauer was among the first to voice objections.

One of the main sources of validation for Halo's approach is a 2008 study posted on its website about transcranial direct current stimulation, or tDCS. The study said that, when applied to the primary motor cortex, tDCS helped in learning a novel motor skill task. Krakauer knew about the study; in fact, he cowrote it. He pointed out that his academic results were not the same as saying you could wear a pair of headphones with arbitrarily positioned nubs and some electrical pulsing to become a better football player. In 2016, a scientist at New York University found that only 10 percent of the electricity applied to the skull actually makes it through to the brain. It was hard to prove that Halo was doing anything beyond a 2-milliamp scalp massage. Arguing ensued. "I've done papers on tDCS and it has been overhyped," said Vincent Walsh, a neuroscientist at University College London. "Mine was one of the seminal papers—I couldn't replicate it."

In 1796, a German anatomist thought he, too, had devised a way to at least understand (though not necessarily to improve) someone's abilities or behaviors with a noninvasive method. Franz Joseph Gall thought he could perceive extraordinary talents not by the physical features of a person's body, but by the anatomy of their brain, which bulged or sagged in certain portions that contributed to the unique behavior. He believed that pickpockets, for instance, had an area on the side of their brain that contributed to a "desire to possess things" and certain zealots had an organ for "religious sentiment" above the eyes that produced large foreheads.

This became known as phrenology. Its reliance on cranial morphology produced scores of crackpot "phrenologists" who speculated about a person's character by rubbing their fingers over the person's head, searching for bumps or indentations. It produced stupid and often dangerous conclusions, and, though it took too long, phrenology eventually became widely discredited. Modern-day brain stimulation tools like tDCS are a much more promising development, particularly for alleviating conditions such as depression or anxiety. But their applications (and corporate spin-offs) might also serve to reinforce how difficult it is to pin a wide array of complex behavioral outcomes on one or even a handful of spots along the cortex. The production of movement, as you may have already guessed and will continue to learn, is not that simple. If it were, we would not still be talking about Hitzig and Fritsch.

The legacy of Gall's ideas about the cortex as a topography for some behaviors is not all bad. It instigated others, like Marie-Jean-Pierre Flourens, Paul Broca and John Hughlings Jackson, to search in different ways for a relationship between area and function. Broca, in 1861, found that patients with speech defects often suffered from damage to their left frontal lobes. And Hughlings Jackson, observing the seizures that afflicted his wife, noticed how the convulsions seemed to spread across the body in predictable patterns. From this, he deduced, correctly, that the motor cortex controlling the right side of the body must be near Broca's area in the left hemisphere, because speech almost always stopped when that side went into seizure.

But as late as 1870, despite many attempts, no one had directly localized movement in animals in the cortex. The French anatomist François Longet had tried to elicit motor responses through the white matter in rabbits and goat kids. "We passed galvanic current through it in all directions without managing to set in motion involuntary muscular contraction," he reported disappointedly in 1842. A year later, Carlo Matteucci found that electric excitation produced nothing from the cerebellum of rabbits. "I can confirm that excitation of the brain lobes, the corpus striatum, and the cerebellum produces no trace of twitching in all free muscles of the body," reported Moritz Schiff in an 1858 textbook on physiology. Perhaps the most respected researcher at the time was Flourens, whose exhaustive investigations into the cortical activity of birds— often removing the cerebrum entirely—left him convinced that the brain's cortical structures might be responsible for sensation and will, but not movement, which had to be generated in some lower portion of the brain stem or spinal cord.

Hitzig, however, had doubts. He has been described as a vain and incorrigible man, faithful to Prussianism. But he knew his way around the inside of a skull. He read of aphasias that robbed patients of certain movements due to the destruction of specific portions of the cerebrum. Hitzig had once even applied galvanic stimulation to the "posterior part of the head" and reported movements of the patients' eyes, "which, judging by their nature, could only have been triggered by direct excitation of the cortical centers." He attempted a similar experiment on rabbits. But he felt the need for a more "definitive resolution." He sought the assistance of a comparative anatomist, Fritsch, to help put the controversy to rest.

With the wire electrodes, they roamed the cerebrum, poking along various points like typists learning a new keyboard. Stimulation of one cortical area moved the dogs' forepaw, while another region moved the hind paw. The center for the neck muscles seemed to lie in the middle of the prefrontal gyrus. The facial nerve could be innervated from the middle part of the supersylvian gyrus. Along other points, the researchers produced definitive contractions in the back, abdomen and tail.

Next, they wanted to see if the dogs would exhibit problems if they destroyed the region that prompted the movement. With the handle of a scalpel, they carefully excavated the "cortical substance," removing a piece "approximately the size of a small lens." They then closed the wound with button sutures. Once the animals recovered, they recorded three observations:

I. When walking the animals placed the right forefoot inappropriately, sometimes more toward the inside, sometimes more toward the outside than the other, and slid slightly outward with this foot, which never happened with the other, so that the animal fell to the ground. No movement was completely absent, however, the right leg was moved somewhat weakly.

II. When standing there were very similar phenomena. Moreover, it occurred that the forepaw was always set down with the back rather than the sole, without the dog noticing it.

III. When sitting on the hindquarters, if both forepaws were placed on the ground, the right foreleg slipped gradually outward until the dog lay completely on its right side.

Though the dogs could still move about and run, often with as much liveliness as before, the researchers continued to notice something slightly amiss. It seemed the dogs "clearly had only a deficient consciousness of the conditions of this limb. The ability to form complete ideas about it had been lost." They suffered from symptoms very similar to those of a disease, one that afflicted not the sensory pathway but the motor. "There was some kind of motor conduction from mind to muscle," the researchers wrote, "while in the conduction from muscle to mind somehow an interruption was present."

Hitzig and Fritsch reported their findings in a paper, *Über die elektrische Erregbarkeit des Grosshirns* ("Electric Excitability of the Cerebrum"), which appeared in a Leipzig medical journal in April 1870. The impact was seismic. William James noted its influence in his *Principles of Psychology* in 1890: The "experiments on dogs' brains fifteen years ago opened the entire subject which we are discussing." For the next six decades, almost every paper on functional localization of the brain contained a reference to Hitzig and Fritsch's seminal work. And copycats proliferated.

In 1874, inspired by the findings of the two Germans, a physician at the Good Samaritan Hospital in Cincinnati, Ohio, named Roberts Bartholow became the first to try applying electrodes to a human cortex, using a 30-year-old Irish servant, Mary Rafferty, as the subject. She had a cancerous ulcer that had eroded part of her skull, opening a two-inch hole through which her brain was exposed. Utilizing this, Bartholow poked around the left posterior lobe, which produced muscular contractions in her right arm and leg, and a forceful head turn to the right. The same phenomena occurred when he applied the electrodes to the right lobe, except the head turn was to the left. Mary, awake and talking throughout the entire procedure, complained of intermittent sensations of tingling. When the strength of the current was increased, she began to cry, then froth at the mouth, and then she suffered a seizure. She lost consciousness and fell comatose for 20 minutes. When finally she awoke, she complained of weakness and vertigo. Mary died shortly after the surgery. Bartholow's nine-page report was published in *The American Journal of the Medical Sciences*, and he was harshly rebuked by the American Medical Association.

On the morning of August 4, 1881, Friedrich Goltz stood inside the already-sweltering Royal Institution in London ready to tear apart everything his intrepid German predecessors, Hitzig and Fritsch, had thought they had proven. He brought with him a dog and a whip. Goltz had excised several portions of the dog's brain in an operation almost a year earlier, a procedure that, he thought, should have left the animal paralyzed and senseless. Instead, it was completely motile, perceptive of external stimuli, and exhibiting only a defect in vivacity and alertness, which Goltz believed established the cortex's role in cognition and intelligence. The fact that the dog could still move its muscles voluntarily, however, beyond simple reflex actions, confirmed to him that "it is not possible to take any muscle by destroying any part of the cerebral cortex," Goltz said. Hitzig and Fritsch, and anyone who believed them, were wrong.

Standing opposite Goltz was David Ferrier, a slight and energetic pupil of Hughlings Jackson, who had years ago set out to confirm Hitzig and Fritsch's work using animals from his own laboratory at the West Riding Pauper Lunatic Asylum, in Yorkshire, England. He assembled pigeons, fish, rats, rabbits, cats, dogs and jackals and, using chloroform to subdue them, managed to expose a much larger circumference of the cortex than the German researchers attempted. Applying a faradic current, stronger than any used in previous experiments, and keeping the electrode on the cortex for longer periods, he could make the animals walk, grab, scratch, blink and flex—complex motions beyond anything seen before.

Ferrier would go on to critique the methods of Hitzig and Fritsch and their "blunted electrodes." Ferrier saw his animals' independent and specific movements as signifying that he alone had stimulated the cortex, while Hitzig and Fritsch had simply misinterpreted their results. "The mere fact that movements result from stimulation of a given part of the hemisphere does not necessarily imply that the same is a motor centre in the proper sense of the term," he wrote. "It will afterwards be shown that the movements which result from stimulation of the regions in question are expressive of sensation, and that the character of the movements furnishes an important index to the nature of the sensation." He challenged the Germans to consider the "character of the movements."

But Goltz, a portly physiologist from the University of Strasbourg, with bulbous cheeks and a thick mustache, took the debate another step further. He vociferously denied the existence of cortical localization as it relates to any movement at all, and he scoffed at the localizationists for their inconsistent results. When the Physiological Section of the 1881 International Medical Congress convened in London, they requested a public discussion between Goltz and Ferrier at the Royal Institution, hoping to end the controversy once and for all.

After Goltz, speaking in German, concluded his long-winded opening statement, denying that the cortex had any role in anything other than cognition and intelligence, it was Ferrier's turn to respond. He didn't bring any animals but said that his monkey "had the motor zone destroyed in its left hemisphere seven months ago," leaving the animal's right side completely paralyzed. "The animal is in every other respect perfectly well," Ferrier said. "And as to its tactile sensibility there is not the slightest sign of impairment." In the afternoon, members of the section were invited to King's College to bear witness to the animals the two men had discussed. Goltz showed his dog around the room, lighting a candle in front of its face to demonstrate that the animal remained completely indifferent, as if unable to recognize it. One attendee blew cigar smoke at the dog to try to get a reaction. When Goltz put the dog inside a box, it moved about and placed its forelegs on the top but could not seem to figure out how to jump out. "This dog must be described as a harmless man," Goltz said. "He is still in possession of all the senses, but he does not know how to utilize the sensory impulses in such an appropriate manner as an uninjured animal."

It was then up to Ferrier to reveal the condition of his animal. From out of a cage, a small monkey wandered toward the center of the room, where Ferrier held pieces of food in his outstretched hand. Its right leg dragged behind its body. It reached for the food with the left arm, as its right arm remained frozen at its side. Spellbound attendees had never seen an animal so accurately show such resemblance to a hemiplegic human. One in particular, Jean-Martin Charcot, known to some as the father of modern neurology, let out an astonished gasp and exclaimed, *"C'est un malade!"* ("It's a patient!")

After the animals were killed, autopsies revealed that Goltz's lesions were far smaller than the ones that had afflicted Ferrier's monkey, sparing what we know now to be the sensory and motor cortex entirely. A separate committee would have to more closely examine the results, but it is hard not to imagine Goltz exiting the meeting with a great deal of embarrassment and, perhaps, contrition. Ferrier, on the other hand, exited the meeting and was almost immediately embroiled in a fantastic scandal provoked by the Victoria Street Society, a London activist group focused on abolishing animal experimentation in England. The Cruelty to Animals Act had been passed five years earlier, requiring annual licenses for researchers wanting to perform studies at specific sites (the law existed until 1986, when it was replaced by the Animals [Scientific Procedures] Act). But Ferrier did not have a license when he spoke to the section, and he was arrested and prosecuted. He ultimately avoided prison time by proving that it was his collaborator, Gerald Yeo, who had actually carried out the surgery, and Yeo was properly licensed.

A third attendee exited the room at King's College with his head spinning with ideas. Just 23 years old, and still an undergraduate at the University of Cambridge, Charles Scott Sherrington would go on to worldwide recognition with *The Integrative Action of the Nervous System*, published in 1906, which would establish him as the singular figure in modern neuroscience who could rise above the rest. The book is dedicated to David Ferrier.

If anything definitive came out of the great localization debate between Goltz and Ferrier, it was that Sherrington was not going to spend much time worried about where movement might be initiated. He trusted Ferrier on that. There was too much to learn about how the spark of a movement proceeded. At Cambridge, his mentor, Walter Gaskell, convinced Sherrington he was wasting his time following Ferrier's example studying the cortex. He should focus on the spinal cord and try to figure things out from the base up. "One could not talk with him long," Sherrington would later say, "without realizing that the cord offered a better point of attack physiologically."

The task of understanding the central nervous system was slightly hindered by the fact that there was not even a clear definition of the nervous cell. Since Jan Evangelista Purkinje had first described a neuron as containing a cellular body and some thread-like extensions in 1837, there was no agreed-upon explanation of how the cells communicated with one another. Neurons did not look like other cells. They were shaped, in a way, like a tree, with a long-rooted trunk sprouting outward from the ovular cell body, then forming a branching network of tendrils, originally called "protoplasmic extensions" and now called "dendrites." Those dendrites somehow fused with the "axis cylinders," now called "axons," of other neurons, enabling the transmission of messages like a microscopic handshake.

Today, we have the technology to enable us to observe the activity of a single neuron. One of the best specimens for this type of study is, oddly, the common household fruit fly, or *Drosophila melanogaster*. I got to see for myself when I dropped into the lab of Rudy Behnia, a neuroscientist at Columbia who studies the neuronal circuits of vision. Fruit flies, when shown under a microscope, are golden brown in color, with large and bright red eyes, and a light pinkish hue to their iridescent wings. Their brains possess only about 100,000 neurons, but they use them to detect motion and light in ways similar to ours. When I visited, Behnia took me into a room in the lab where a student was working in near-pitch darkness, her face illuminated only by the glow of her computer screen.

On the screen was a window displaying faint green twinkles of light against a dark background. It looked like an image of the electrical grid of the United States at night. At certain instances, some of the twinkles grew brighter and others dimmed, as if cities on the map were blinking their lights. These activations and deactivations corresponded to the fly's neural activity, in real time. At that exact moment, the student was switching on and off a UV light in front of the face of a fly that was fastened to a prop just a few feet away. She was studying how the fly's neurons interact.

In humans, the messages transmitted by our neurons can travel anywhere from a few micrometers to nearly a full meter, in the superhighway axon located in the femoral nerve in our leg. Neurons make up anywhere from 40 to 50 percent of the cells in our central nervous system; the rest are glial cells, which derive their name from the Greek word for "glue." Glia are believed to play some role in neurotransmission, but they don't possess the communicative properties of our neurons. There are about 1,000 types of neurons. Movement relies primarily on three: sensory neurons, motor neurons and a kind of intermediary cell called "interneurons," which serve to modulate the other two. All three interact not in the brain, but in the spinal cord, and in particular, the H-shaped core region of gray matter running vertically up the length of our back. The upright columns of the H are called "horns," and they are divided into two sections: dorsal (or posterior) horns and ventral (anterior) horns. The dorsal horns contain sensory neurons gathering all the afferent signals sent from everywhere in the body. The ventral horns have the motor neurons. They are the output messengers, informing the muscles of what the brain wants them to do.

The spinal cord itself is also divided into four major regions:

cervical, thoracic, lumbar and sacral. Very few sensory axons enter the cord at the lowest (sacral) level, but the number increases progressively up the spine. The sizes of the horns also change. Our spine widens at the top and toward the bottom, where the input and output of commands to and from the arms and legs is more robust than what is needed to move the torso. The central nervous system (brain and spine) is constructed to generate our capacity for movement.

In 1873, Italian physician Camillo Golgi, working in his kitchen, discovered that a combination of potassium dichromate and silver nitrate could create a silvery stain on living neurons that would allow scientists to visualize the full nerve cell through a microscope. Golgi—and others—thought neurons interacted collectively, through physical linkages, as though they were part of a connected loop or a reticulate net. They transmitted messages up and down the spine and throughout the body so quickly, and appeared in the brain so tightly packed together, there seemed no other explanation. "Nerve cells, instead of working individually, act together," Golgi said. Yet there was a discrepancy about how the structure was actually composed. Were the cells fastened to each other like a network of blood vessels? Or were they stacked together like bricks? Santiago Ramón y Cajal designed his own staining method, which allowed him to better visualize what was happening at the molecular level. He did not see any connection between cells at all; rather, the axons seemed to end in "baskets" that never physically convened with other dendrites. Golgi and Cajal, from opposing camps of nerve theory, would never see eye to eye.*

---

* The two men famously shared the 1906 Nobel Prize for their contributions to cell theory, even though, by that point, Golgi's nerve net hypothesis had been widely dismissed.

When Sherrington began to investigate the nervous system, he started with a more observable behavior. At the time, the section in Michael Foster's third edition of *A Text-Book of Physiology* called "Reflex Actions" was only five and a half pages. The book was 1,292 pages. Sherrington could not accept that that was all there was to know about the phenomenon of reflexes.

When you go to the doctor's office, and she strikes your kneecap with a hammer, the sudden knee-jerk reflex is familiar to all of us. It is the same reaction that throws the muscles of the calf into contraction by tapping the Achilles tendon. But it was Sherrington who figured out the chain of sensory and motor components involved, the pathway to the spinal cord, and the methods with which the central nervous system could modify or even inhibit it. Tapping the patellar tendon with a hammer actually pulls on the quadriceps femoris, which stretches the muscle in your thigh. This information is then relayed along sensory pathways snaking throughout our limbs that empty into the spinal cord, which signals to the motor neurons there that they should contract the quadriceps. Simple.

But that's still not the whole story. In the spinal cord, the interneurons are also activated. These will act as an inhibitory force on the motor neurons that feed commands out to the hamstring muscles. With the interneurons engaged, the quadriceps can contract unopposed, because the hamstring has been effectively silenced. The familiar, irrepressible leg kick is produced—the result of a two-way communication between the musculature and the spinal cord.

This sequence is among the simplest behaviors our bodies can

Nonetheless, Golgi spent the majority of his acceptance speech defending "interstitial nerve nets" and assailing Cajal, as attendees "looked at the speaker in stupefaction."

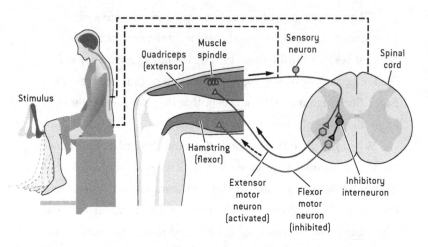

The knee-jerk reflex, described by Sherrington in 1892, involves a monosynaptic circuit of signals passed from sensory neurons to motor neurons through the spinal cord. A third class of neurons, interneurons, act as an inhibitory agent, preventing the hamstring from contracting and allowing the quadriceps to shoot out unopposed. This simple behavior is an example of the reflex arc, which allows us to maintain our balance and homeostasis. The associated movement occurs without the involvement of the brain.

produce—a monosynaptic reflex that, Sherrington would realize, does not even involve the brain until the movement is concluded. Yet it told him a sophisticated narrative: the so-called reflex arc, the feed-forward loop of sensory-to-motor connections that trigger our everyday actions. This loop is always in a state of vigilance, ready to "jerk" into action, "much as an angler by the 'feel' of his line is kept aware, ready to strike," Sherrington would write.

He was profoundly evocative like that. Though his papers were exhaustively detailed, and his sentences could drag on, he sprinkled his reporting with a lucid and atypically panoramic perspective. William C. Gibson, a former student, called Sherrington an "interpreter." He was also, in many ways, a philosopher. Sherrington remarked in 1933: "Inside the animal's form sits the brain, its work broadly to increase the animal's grip on the world about it, and hardly less the grip of the external world upon the animal. Grown up with the animal, it fits the motor mechanism of the animal as much as a key fits its lock. A question the curious ages never fail to ask is, who turns the key?"

Not surprisingly, Sherrington wrote poetry his whole life. Born near the train tracks in Islington, though he was never clear about the year, Sherrington lost his father, James, who died when he was young. His mother remarried a scholar, Dr. Caleb Rose, who imbued Sherrington with an appreciation for antiquities. At Queen Elizabeth's Grammar School, in Ipswich, he was tutored by a promising young poet named Thomas Ashe and was soon composing verses of his own. As a first-year resident at St. Thomas' Hospital, homesick, he sent doleful poems back to his mother:

Dear sleeping eyes that dreamed perhaps of me,
And drew from memory's place,
Forgetful how the months elapse,
Some homely vision and my face.

Sherrington was small, only five foot six, but he rowed and joined the rugby team at Cambridge, and he was particularly fond of a somewhat reckless activity known as riding the "bone-shaker"— a bike without brakes. He once visited Bordeaux, left penniless, and found his way back to England on a wine boat from the Gironde. At a bookstore in Cambridge, he discovered a rare first-edition copy of Keats, which he purchased for sixpence. Fifty years later, he would sell it back to the same bookstore for 100 pounds. He revered Italian artwork and Shakespeare, canoed on the Thames and petitioned Oxford to allow women into the medical school. He regularly discussed theatre with Bernard Shaw, argued with Pavlov and once dined with the Russian tsar. He also nearly died, in Vancouver, upon opening the elevator doors of a still-renovating hotel and narrowly escaped plummeting seven stories. "It's a new country," he remarked, dryly. "Not everything is finished yet."

In 1894, Cajal actually visited Sherrington at his home for several weeks. Sherrington noticed that Cajal kept the door to his guest room locked, so no one could disturb his carefully prepared slides of the nerve fibers he planned to present. He invited Sherrington one night to take a look, and, in a manner that struck Sherrington as odd, he spoke of the specimens as if they were still living. A nerve cell, he would remark, "groped to find another." It was as though, through the microscope, Cajal entered another world, one that Sherrington wanted to join. "There was clearly an element of greatness in him," Sherrington would later say.

Cajal's gaze rarely strayed above the ocular of his microscope. Sherrington, by contrast, could not help but assess the problem of the nervous system more globally. "To move things is all mankind can do," he said. "And for such the sole executant is muscle, whether in whispering a syllable or in felling a forest."

The body contains both agonistic and antagonistic muscles, working opposite but not against each other, like the tips of a scale. For example, when we flex our arm at the elbow, the agonist biceps muscle contracts while the antagonist triceps muscle relaxes. In the knee-jerk reflex, the agonist quadriceps muscle stretches when the antagonist hamstring is impeded. This yin-yang relationship had been observed, but not fully understood. Galen thought the antagonist muscle simply lay dormant. Descartes had supposed that muscle antagonism was more of an "active" process than consequential, but the nature of the relationship was unclear. Early in the nineteenth century, Charles Bell wrote of nerves that "relaxed" the muscles as other nerves excited them. But Sherrington wanted to understand more.

To him it seemed odd when, after he had surgically removed the cerebral hemispheres of a monkey, and the unconscious animal began to breathe normally again, a "peculiar rigidity" ensued, locking up certain joints well before the onset of rigor mortis. The knees were stiffly extended, and the tail tightened into a pole. The head retracted and the chin jutted upward. In the case of the rabbit, the knees of the hind legs freeze, while the wrists and ankles remain flaccid. And when he tried this with a cat, removing only one of the hemispheres this time, an even odder phenomenon occurred. Both legs opposite the ablated hemisphere stiffened, while those opposite the intact hemisphere did not.

Sherrington found that he could terminate the rigidity simply by snipping the afferent root inside the spinal cord. Why would an interruption of the sensory pathway have such an effect? Sherrington wondered. "Decerebrate rigidity," as he called it, "seems therefore in some way dependent on integrity of the afferent paths of the limbs," he wrote. But there were other ways to impede the onset of rigidity. Strong electrical excitation of points along the cerebellum of monkeys, for instance, could cause relaxation of the rigid neck and tail muscles. And when he applied currents to the

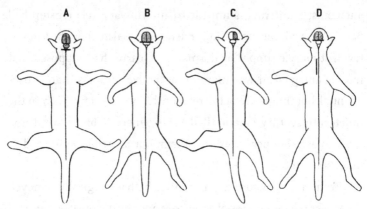

Sherrington described four postures of a cat following decerebration, or surgically transecting the animal's midbrain. A) Normal position of the cat after a lower transection. B) The "peculiar rigidity" after transection of the midbrain. C) Removal of one of the hemispheres causes stiffening to only the opposing limbs. D) Stiffening of the limb ceases when the sensory pathway in the spinal cord is cut.

pinna of one ear of the cat, the body changed its formation entirely. There ensued a turn of the head, a flexion of the elbow and a relaxation of the erect tail.

Such a reflex, Sherrington noted, "gives the impression of an attempt to escape from the irritation." If the opposite pinna was irritated, the same thing happened toward the other direction. He tried this with various excitation points along the paws, all to similar effect, all "suggestive of a *purpose.*" He noted relationships of certain muscle groups to one another, called "reciprocal innervation." When walking, certain leg muscles extend while others flex, enabling us to maintain balance. Concomitant with that, the neck muscles keep our heads poised, and arm muscles allow our limbs to swing rhythmically. When Sherrington saw this coordinated action in the decerebrate animals, it told him an additional story. The

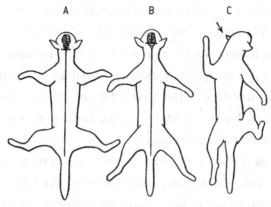

A more pronounced reaction was elicited when Sherrington applied stimulation to the cat's ear. He thought these movements seemed "suggestive of purpose," though the cat had no cerebral connection. A) Normal position after a lower transection. B) Decerebrate rigidity after midbrain transection. C) Change of position evoked by stimulation of left pinna.

reflex and inhibition action produced a motor pattern that resembled simple movements, like walking or turning. Yet they took place without an ingredient thought to be essential to the production of movement. The animals had no brain.

He would go on to produce and discover other characteristics of reflex actions in dogs and cats with spinal lesions. When he rubbed the skin of a dog between the shoulders, the reflexive movement resembled a "shake," as if the brainless dog was trying to rid itself of whatever was bothering its skin. When he pricked the skin of the neck, the dog's movement quite clearly resembled a "scratch." The scratching can follow the source of the stimulus as it ventures over the skin, and it can continue for long after the stimulus is removed, all while maintaining a standing posture, and all without the need of a conscious thought. The paths of the scratch reflex, Sherrington realized, begin and end in the cord. The revelations helped Sherrington to think differently about reflexes. Prior to his work, four interacting factors were thought to determine coordinated reflex movements: (1) the character and (2) the intensity of the sensory impulse, (3) the location of the stimulus, and (4) the solid condition of the spinal cord. He had shown that, in fact, it took very little for reflex movements to unspool, thanks to the reciprocal innervation of antagonistic muscles.

From there, one could produce a whole range of simple to even relatively sophisticated movements. "The main secret of nervous co-ordination," he wrote, "lies in the compounding of reflexes." The humble mechanism of excitation and inhibition functioned at a spinal level, without the higher processes of the cortex, and could still produce natural and archetypal examples of motions. Reciprocal innervation appeared to be a basic building block in motor coordination, one that could be linked together with more building blocks, forming a reaction chain that resulted in something more than just a jerk of the leg. It formed the basis for the machinery of everyday action. Reflex action, Sherrington would later say, "is independent of consciousness even at first occurrence. It does not emanate from the 'ego.' The reflex is innate and inherited."

Sherrington's remarks on the mechanisms of coordination, composed in 14 "notes" written between 1893 and 1909, "opened the way to the further advance from the simple to the complex," Edgar Adrian, who would share the Nobel Prize with Sherrington in 1932, once wrote. "It was the clue to the whole system of traffic control in the spinal cord and throughout the central pathways." In his diligent and exacting experimental style, publishing often without a single coauthor, Sherrington had outlined the gross rules that governed the function and mechanisms of basic movement, and they had nothing to do with the brain. Over the course of a decade, he would also go on to painstakingly map the sensory nerves from the skin to the spinal cord, as well as the motor nerves going from the spinal cord back out to the muscles. His findings produced the first complete picture of the reflex functions of the spinal cord. In an 1897 lecture, he likened it to a "funnel": "The wide entrant mouth of which is represented by sensory nerves, the narrow end of exit by the spinal motor roots to the musculature."

Sherrington had a detailed picture of reflex motor behavior when he was invited to speak as Yale's second Silliman lecturer, in April 1904. The title of the lecture series was intended to be on the "wisdom and goodness of God as manifested in the natural and moral world." But Sherrington clearly did not travel to New Haven to deliver a religious sermon. At age 46, having worked more than half his life in the laboratory, he hoped to paint a clearer picture of what he considered the "integrative action of the nervous system." In anticipation of his talk, Yale prepared a dinner menu with items labeled "Some of Nature's First Experiments with the Synaptic System on the Half Shell"; "Metameres of Skeletal Musculature from a Higher Vertebrate"; and "Saccharine Stimuli of Trigeminus" along with "Caffeine for Vaso-Motor Reactions."

His 10 lectures over the course of 13 days inside North Sheffield Hall would eventually be composed in a book published in 1906, turning into a classic overnight. It was in these lectures that he introduced his hypothesis about the major architectural feature of the nervous system: the "final common path" of the reflex chain. When a sensory neuron sends a signal toward the spinal cord, it joins a large pool of other signals from all across the body. It is like stepping out of a cab and into Times Square. Immediately, you are swarmed by people. You are trying to get to a specific street, but throngs of tourists—all with different intentions—are jamming the roadways. Some are pulling you forward (excitatory) while others are holding you back (inhibitory). Thankfully, there is a traffic cop who can sort through the crowd and signal you on your way down the correct path. That would be the "final common path," the path that leads to the musculature to trigger a movement. A specific type of motor neuron serves as the traffic cop, integrating all the incoming streams of impulses and appropriating their final direction. That pathway might be determined by whether there are more excitatory or inhibitory tourists at that particular convergence point. But ultimately the motor neuron makes the call. "It is the sole path which all impulses, no matter whence they come, must travel if they are to act on the muscle fibers," he said.

The decision by the traffic cop, ultimately, is reliant upon the "synapse," the microscopic gap between the axonal terminus of one neuron and the dendritic tendrils of another, *the space between*, where signals are communicated and transferred. The electron microscope had not arrived yet, so no one could physically distinguish the tiny gap between neurons that Cajal had for years insisted was there. But Sherrington, siding with his old peculiar friend, not only agreed with Cajal's "doctrine" of nerve cell independence but also deciphered the way it existed. In 1898, he had soberly surmised that neurons must correspond with each other across some form of a channel, initially calling it a "synapsis," after the Greek word "to clasp." The synapse must delay the process of transmission—he had deciphered this by noticing an additional latency between how fast signals traveled in the reflex arc compared with the fundamental speed of signals traveling through a nerve—as well as somehow route the signals to the proper destination, like a telephone exchange. The synapse was responsible for guiding the signal onto its final common path.

As he spoke to his rapt audience in 1904, Sherrington laid out a comprehensive model for the entirety of reflex action, from synapse to muscle. His masterwork encompassed evolution, anatomy, histology and physiology, unified by the research on the reflex arc he had contributed over the course of 25 years. From that point forward, he would be known as the "architect of the nervous system," for recognizing and detailing the floor plan that orchestrates coordinated action in any vertebrate organism. Some of the attendees might not have known what they were listening to at the time. "I understood all the words," one told the *Yale Alumni Weekly*, "but I sat there for an hour, and I did not know a thing the man had said." But by the time the book was published, there was no mistaking Sherrington's impact. "In physiology," reviewer F.M.R. Walshe wrote in 1947, "it holds a position similar to that of Newton's *Principia* in physics."

I looked for whatever I could find on Sherrington, including visiting collections of his correspondence at the archives of the Royal Society and the Wellcome Trust in London. But one day, I happened to be at the New York Public Library, not far from where I live, when I found a piece of his writing that I had never before seen referenced. It was a modest foreword to a brochure-size book from 1923 titled *The Brain and Golf* by C. W. Bailey. It was buried in the stacks below Bryant Park.

I waited almost an hour for someone to retrieve it for me. The faded jacket was the color of dry fescue. The foreword, I was surprised to find, was only about 400 words. In academic papers, Sherrington has written sentences nearly as long. I have no idea why such a luminary—not yet a Nobel laureate but still one of the most esteemed British scientists of his day—would have made time to write it, other than by the grace of his own gallantry and humility. After all, this was a man who, on the night he learned he was awarded a Nobel, told nobody and went to dinner at the dining club at Merton College. "The award comes as a complete surprise," Sherrington wrote in a letter to his former mentor, Edward Albert Sharpey-Schafer.

Likewise, his blurb for Bailey was unmistakably Sherrington, praising the author for such an "attractively written little volume" that "shows incidentally that to devote a page to science is not necessarily to be dull." But I was interested in hearing Sherrington discuss "skill"—the broader manifestation of the muscles and joints and nerves and axons he had so expertly analyzed with furious precision for so many years. I knew, as he grew older, and severe rheumatoid arthritis wracked his body, forcing him to frequent mineral baths and early retirement homes, Sherrington's focus drifted away from the singular subject he had obsessed about for more than half a century. "The reflex was a very useful idea but it has served its purpose," he told a former student, Russell Brain, late in life. "What the reflex does is so banal. You don't think that what we are doing now is reflex, do you? No, no, no." He began to think more deeply about the relationship between mind and brain, and the role that consciousness played in the control of advanced human movements. Judith P. Swazey, writing about Sherrington in 1969, contended that he was probably not a philosophical dualist by choice. "It was a position he felt constrained to adopt because the sciences of the day offered no evidence, no means, for bridging the gap between mind and brain."

Reflexes, he would write in *Man on His Nature*, published in 1940, started the "mind on its road to recognizability." But the mind only concerns itself with the "why" of a motor act, not the "how." When eating a bite of food, your attention can be focused on it: chewing, moistening with saliva, savoring the good taste or discarding the bad. All this would seem "at will." But once the food is swallowed, it is no longer of mind, even though the muscles and nerves continue to engage with it, passing the food along toward digestion. "One use of mind in the individual is to control and modify the individual's motor act," he wrote.

Where the mind and the body do seem to interact, though, is in the cortex. Sherrington called it the "roof-brain," and its development distinguished man from beast. "The roof-brain component increases the finesse, skill, adaptability and specificity of the motor act," he said. A dog without its cortex could still stand and walk, but it could no longer be adapted to a special purpose or trained on a given skill. It could walk to but not greet its master. "The roof-brain alters the character of the motor act from one of generality of purpose to one of narrowed and specific purpose," he said. "It is just as if the body and its finite mind had become one!"

With that in mind, when I got to read Sherrington's short passage in Bailey's book, I recognized, for the first time, his appreciation for the complete picture of movement, cortex and all. I recognized, too, those shades of John Krakauer, awestruck at the motoric sophistication of a Federer or Curry.

Here was Sherrington, in his modest way, dishing his own thoughts on athletic intelligence, as if it was a little secret he was dying to whisper to whoever might listen. He was out from the constraints of his laboratory. The brakes of his bicycle had been stripped.

"The difficulty of expressing in words the experiences which go toward the acquirement of muscular skill is great," Sherrington wrote. "Our language has, so to say, allotted all its vocabulary to other senses and experiences than the muscular, especially to sight and hearing.

"But that may not imply that even in our most deferential attitude towards the 'intelligentsia' we can afford to turn our backs upon training in experiences which can be rich although embarrassingly inarticulate. Enthusiasm and respect for athletic games were a feature of ancient Greek culture in its hey-day.

"The acquisition of golf skill is an exercise which may be undertaken in that spirit here and now, where and when, more than ever, it is needed."

5.
THE NEUROTECH SPACE

OUT OF THE LAB

Hello, everybody. We're deCervo. Imagine you're a baseball player, you're a hitter and you have to decide whether to swing at an incoming pitch in under half a second. We're able to measure when you're making those really quick decisions, and how that relates to your skill on the field."

Jordan's voice quavered at the MIT Sloan Sports Analytics Conference in Boston. He was nervous, teetering on his heels beneath heavy spotlights on a small stage before a quiet roomful of potential investors, industry leaders, sports executives and college students in suits and backpacks who thought they should be future sports executives. It was a clear and gelid day in early March 2017. DeCervo had been invited as one of 24 companies to participate in a startup competition for a prize of $5,000. The price of entry, Jason noted, was $3,000. But there were other benefits to being there. They had a booth in Room 311 of the palatial Hynes Convention

Center. They had walk-by access to more than 4,000 data-and-tech-savvy attendees, including Mark Cuban, "Moneyball" patriarch Billy Beane and NBA commissioner Adam Silver. And they had "five good minutes" to pitch their company to a panel of well-heeled entrepreneurial judges, as Jordan was doing now. "The underlying principle behind this is that expertise literally changes how our brains work," he said, reciting lines he has memorized by now. "Understanding how you can measure these changes in the brain, and how that relates to on-field performance, provides potential value for teams."

Jason stepped onstage and introduced what he called the "uHIT Platform," the full oeuvre of options that deCervo had carefully crafted to distribute their product to the masses. It had been burnished by the insight they had attained from two years of working, or otherwise attempting to work, with Major League teams. Their business acumen had indeed improved. Jason spoke of "permeation" into the marketplace and a hierarchy of data plans. At this conference—a two-day sports-geek mecca, cofounded by the Houston Rockets' general manager Daryl Morey and sponsored by ESPN—they were ready to unveil their latest mobile and virtual editions. They had brought two tablet computers and the 40-inch Hitachi television screen from Jason's living room to set up at their booth for anyone willing to try. Ever the showman, Jason also brought a rubber home plate and masking tape to form a batter's box on the carpet. "It's gimmicky," Jordan protested. "Yes, it is!" Jason said.

Their booth was a hit. Crowds gathered to watch men and women in suits try to time the incoming fastballs and sliders that seemed to approach at impossible speeds. "I didn't realize this was a spectator sport!" Jason crowed. The game itself had matured: Instead of a green dot on a blank screen, a dazzling CGI baseball stadium gleamed under azure skies and cottony cumulus clouds; there were mower streaks on a verdant field and deCervo banners on the outfield walls. The looming scoreboard in center field lit up green for correct responses, red for incorrect. They had preloaded the release points and pitch repertoires of real aces, like Clayton Kershaw, Jake Arrieta and Zack Greinke. They had also created a contest: Whoever finished with the highest scores in the Strike Recognition and Pitch Recognition games received a black, fitted deCervo baseball cap. Whoever finished with the highest totals in both games won box seats to a Mets game. When a member of the Chicago Cubs business operations team came by, they downplayed the latter reward. When a man from Germany came by, Jason addressed him in German. When a man from Québec came by, Jason greeted him with *"Enchanté."* Over the course of two days, they received more than 60 visitors. Shane Battier, the two-time NBA champion, even ambled over to test his virtual batting skills. Personnel from baseball, basketball, football and soccer clubs gave Jason and Jordan their cards; a member of the Australian Institute of Sport asked how deCervo could be used for cricket. The NBA expressed interest in working with deCervo as a training tool for referees.

Though pleased with the universal appeal of the game itself, Jason did not want to take any chances. His girlfriend, Celine, a perfumer, had gifted him a small spray bottle labeled "deCervo Grass #1" before he left. She informed him that men are instinctively attracted by scents from the outdoors. Without mentioning anything to Jordan, Jason had been surreptitiously spritzing the smell of a mown lawn around his booth, hoping to lure executives with a whiff of baseball's future.

When Jordan was born on December 2, 1984, at New York Hospital, his parents did not have to go far. "Just took me across the street," Jordan said. They lived in a postwar high-rise on 72nd Street and York Avenue. He grew deaf to the ambulance sirens. His bedroom window, growing up, faced Sotheby's. "They never turn off their lights," Jordan said. In other words: "I can sleep through anything."

He was telling me about this, laughing, from the farmhouse he and his wife, Izzie, had recently purchased off a dirt road in Bedford Corners, New York, surrounded by thick deciduous forest, about 15 miles from the Hudson. Their four-acre property rests atop a gently sloping meadow, with five apple trees, stone outcroppings, and clear views to the north and east across the bedroom communities of Fairfield County. Some nights, Jordan said, they can even see the twinkling lights of New Haven. The only sound I heard when I left was the yelping of a few coyotes.

As a kid, Jordan was so quiet his parents would joke that he might be mute. His friends nicknamed him "Doc," because when he talked, he usually did so about going into medicine. This satisfied his mother, Vicki, who said that she ate salmon almost exclusively during her pregnancy because she had read that it boosted the child's intelligence. Indeed, Jordan was smart and intellectually curious, with an eye for aesthetics. He loved the tulips on Park Avenue. He also often accompanied Vicki on trips to the Hospital for Special Surgery, where she rehabbed from three scoliosis surgeries that had derailed her career as a dancer and dance instructor. The scoliosis was genetic; Jordan's older sister, Brooke, developed it, and then Jordan did, too. For 23 hours a day for two years (age 11 to 13), Jordan wore a stiff plastic brace around his midsection to straighten his spine. The brace worked—Jordan's spine angle returned to zero degrees. Then he grew seven inches in a year. The curve grew with it. Though he never wore the brace again, he is still troubled by back pain. He stands six foot one, but Vicki says he should really be more like six three.

In spite of the back condition, Jordan did not stop playing and excelling at sports, especially golf, which he picked up from his father, David, who played at Rutgers and became one of the best amateur players in the region. Their relationship, though, was strained. On December 21, 2001, shortly before Jordan graduated from Horace Mann School and entered Columbia as an undergraduate, David, an attorney, was charged with one count of wire fraud, to which he pleaded guilty, and was sentenced to two years in prison. His parents divorced, while David was disbarred and moved to Florida. Jordan went through periods of confusion, disgust, loss, rage and, eventually, reconciliation. "It shaped me as a person," he said. Jordan never brought it up with me, but also didn't shy away when I asked. He and his father do talk more now, he said.

One summer evening, the kind of languid summer evening when the air feels so pure and the sun recedes so slowly that you might look at your watch thinking it is finally time to head inside for dinner, only it is quarter to 10, and the lateness of the moment surprises you, on this kind of evening Jordan and I were sitting outside his home and drinking wine and watching his dogs, Girtie and Augie, tear after a deer they must have sensed somewhere down along the shadowy edge of the meadow. The outdoor air smelled a bit like deCervo Grass #1. There were crumbly piles of dirt burrowing up from various cracks in the patio. They looked like low-lying anthills. "Wasps," Jordan said. "They come out of those nests in the ground in the summer and kill the cicadas." He pointed out his apple trees, growing nicely on the lawn. He pruned the trees and picked the cherry tomatoes from the garden to prepare as a salad with oregano and red pepper flakes. "He's the outdoor wife," Izzie teased.

When we went inside, Jordan took me to his office to show me some of the progress he and Jason had made with deCervo, including the uHIT Platform, which was now much more user-friendly and streamlined. On his laptop, he also showed me an example of their next, arguably their greatest, advancement, something he did not mention at the Sloan Conference, not even to potential investors: real video. Instead of a simulation, there was actual footage, taken from the catcher's vantage point, of a professional pitcher throwing fastballs and sliders. Teams had been asking for this, he said. "I've never been more confident," Jordan said.

He paused a second. "We've obviously moved away from EEG. I didn't want to. But it's a no-brainer at this point. I worried we would be moving away from our strengths and become like a video game company. If we became just that, we wouldn't be competitive. Because we're not video game developers."

"Then why are you still so confident?" I asked.

"It's still developing in a way that is still founded in neuroscience," he said. He added, "We have so much data right now that we haven't even looked into. Once we have the time to take a deep breath, we can really dive into it and start asking questions."

I asked him what questions, and Jordan's scientific curiosity rekindled. He mentioned an idea related to using the system to create "the perfect batter." They could simulate how a robot might reach 100 percent accuracy at the shortest possible instant, say 100 milliseconds. That might be unattainable for a human. But deCervo could possibly work backward from there, determining when the infallibility begins to crack. "Now you can start peeling away at an understanding why batters might not be as good as a 'perfect batter,'" Jordan said. "What and why are certain pitches fooling them?

"We have these ideas," he continued. "We have these visions. We just don't currently have the time. But a lot of this stuff would never happen in academia. We could never get live videos from teams to fit what we need. We could never get this many pro players to use EEG."

At the moment, Jordan and Jason were still trying to understand why they were hitting roadblocks along the path to full buy-in from Major League clubs. Baseball might have been the first league to wave the era of advanced analytics around third, but presuming every team is by nature forward-thinking is a misjudgment. Organizations are generally eager to test but averse to adopt. The view is that they can afford to be dubious but not afford to be duped. Franchises are as wary about getting left behind as they are about being the first to step into the breach. They can also be paranoid. Sometimes this is for good reason. A former scouting director for the St. Louis Cardinals, Chris Correa, was sentenced to nearly four years in federal prison in 2016 for hacking into the player-personnel database of the Houston Astros. Teams will often ask companies like deCervo to conform to certain limitations, based on a desire not to rock the entire organizational boat. But this can alter the product.

In deCervo's case, it was starting to hamstring their push for scientific advancement.

Early on, teams seemed to regard deCervo as a novel curiosity—a cool, futuristic technology that intuitively made sense. "It opens up a new lens to evaluate hitters, which is something we're always interested in," an American League assistant GM gushed to me. "It's an emerging field with some promise, and we want to have our pulse on what's going on in those areas." As the months wore on, though, it became clear from speaking to teams that they did not have much beyond a foggy understanding of what "neural data" could provide. "My front office goes, 'We want neural,'" one National League performance coordinator said with a sigh. "I'm like, you just explained everything in the human body essentially." Others struggled to figure out how to weigh the costs of experimenting with a product that required multiple EEG sessions. "I think we have to be careful about knowing more and more about less and less," the AL executive said. "Time is precious. Practice time is precious. Am I going to tell guys they should be doing a half hour of this training instead of a half hour of something else? I want to make sure there's a good case for doing that, not just, oh wow, this is interesting, I now know his reaction time down to the nanosecond."

The biggest barrier to entry, though, seemed to be the sport's natural conservatism. This can manifest itself within the game, where flipping a bat after a home run still warrants a beaning in the next plate appearance, and outside the game. "They still look like scientists," one observer, a former professional soccer player, remarked to me at Sloan. It was not meant as a compliment. His fundamental point was that pro-ball execs would be unlikely to read the academic esoterica that may or may not validate a product like deCervo wanted to sell to them. Baseball had flooded its front offices with Wall Street exiles and Ivy Leaguers with MBAs. But it remained suspicious of the scientists. "We actually have to convince them we're less scientific," Jason says. "Science is scary. It's more important for us to speak baseball than neuroscience."

In spite of the language barrier, there were still teams at least showing interest in incorporating cutting-edge neural science into their organizational offering. And as long as they still had unique access to the brains of professional athletes, Jason and Jordan were invested. Frank's team had flown Jason and Jordan twice to their Arizona facility to test players with EEG at the beginning of the season and, again, at the Hilton in August. They had also been the first team to implement the full gamut of deCervo's strictly behavioral products: uHIT Mobile (on iPads) and uHIT Virtual, which they projected onto an eight-foot screen fastened to the back fence of a cage inside the outdoor batting tunnels at the player development complex. A dusty ProBook rested on a wobbly wooden stool 10 feet away, above an LCD projector strung with cobwebs. There was little doubt the players felt more comfortable training their pitch recognition skills inside the cage. With a handheld controller to trigger their responses, they could even feign a batting stance, enhancing the realism of the scene. It looked like a fully immersive video game. It looked more or less like Jason had once described his vision: "DeCervo can fit right into their normal baseball routine," he said. "So a player can go from hitting off a tee to soft toss to uHIT."

But, to Frank, it was still a bit unclear what exactly the players, who had been asked to incorporate the game into their daily routines, were training, or how these behavioral tasks aligned with the neural profiles they had been creating. "I have scientific curiosity," he said. "But in my job here, if I can't improve something, there's no sense in measuring it." He didn't seem to be thinking about deCervo's potential as a scouting tool, but I didn't pursue it.

One afternoon, Frank sat with Jason and Jordan around a dark oak table in an air-conditioned conference room on the second level of the training facility, as Jordan flipped through a PowerPoint presentation. Scattered around them were coasters featuring All-Star Game program covers. There was a bat rack inside the built-in media console. On one of the walls was a framed photograph of a coach eyeballing a hitter inside a cage, head on his chin in deep concentration, and I could not help but note the irony. The coach was sizing up physical characteristics; they were now sizing up how much those physical traits were betraying them. But a long weekend had obviously taken a mental toll on Jason and Jordan, who led a meandering talk that was starting to give Frank a headache.

"Wait a minute, wait a minute," Frank interjected. "An incorrect 'Go' is when you swung when you shouldn't have?"

Jordan: "You swung when you shouldn't have."

"But those are the same pitches that you're going to a correct 'No-Go' on."

"No, no. An incorrect 'Go' is when you didn't swing on one where you should've swung. And a correct 'No-Go' is where you didn't swing on one that you shouldn't have swung."

Frank paused. "Okay, so an incorrect 'Go' is when you didn't go when you should have."

"Yes."

"Go back to your first slide."

Jordan flipped back. Jason cut in: "We define it as 'Go,' so they responded, and incorrect because it was incorrect."

Frank: "But according to your first slide, this is the same pitch, so as one goes up the other has to go down."

Jordan: "You're right. Oh yeah, you're right."

After the Abbott and Costello routine, Frank raised some valid concerns. He holds a master's in biomechanics and a PhD in sports medicine, and he has been a certified athletic trainer since the early 1990s, which gives him some latitude for objections. He was worried mainly about the players reaching plateaus, and whether or not the platforms uHIT Mobile and uHIT Virtual were enticing enough to get the organization entirely on board for uHIT Neural. To him, there was reason to proceed with the EEG assessment without the other pair. And without uHIT Neural, deCervo would start to look like Jordan's biggest fear: They were just another video game company.

"How are you using it now?" he asked. "How is it usable in a performance setting? *Is* it usable in a performance setting?"

"Those statistics are more predictive," Jordan said. "In our previous work with other teams, we've seen that neural data predicts better than the behavioral."

"So if we look back at the beginning of the neural data," Frank says, "does that give us any clues as to who would be better performers?"

Jordan: "We want to get to the data first before we start cherry-picking and saying X, Y, Z—"

"That's fair," Frank interrupted, "but those questions are important. They give me the opportunity to say we need this neural data. If we're seeing these trends in their neural data when we've tested them, so maybe we need to train them in this way or that way. Or, we test these players and this is a player that, honestly, we're probably not going to need to work with as much, because there's not that capability there to improve. I hope that's not the case with anybody, but to be honest, it is. There are some people that don't have those neurological abilities to advance."

Jason and Jordan both shifted uneasily in their seats. They were a long way from the lab. Academia can move at a glacial pace. The comeback to that response, if posed in a more academic setting, would require reams of evidence, dozens of citations and months of painstaking peer review. And after just two neural sessions, it was hard for them to stomach the idea that a team could decide about a player's future based on fewer than 500 pitches in a laptop simulation. But sports teams want definitive answers right away. If not definitive, then anything they can use to convince ownership it is worth the cost.

Jordan carefully framed what he planned to say next. "This data allows us to begin answering a lot of these questions," he said. "There are differences among players between the time of their decision and their response. Knowing that difference is information you guys can use."

He mentioned visual occlusion as a training method to potentially incorporate with the uHIT in the cage—blocking out the view of the ball at a certain point as it approaches and forcing the batter to respond based on his best guess of what the full pitch might look like. Visual occlusion is becoming more readily incorporated into practice settings as a method of training prediction and decision-making. The timing for when the ball should be blacked out has generally been an educated guess. Now, with a player's precise neural data, a team can see exactly when his decision is being made and time the occlusion to either hasten or delay that decision.

"It's knowing what the prescription should be, based off the diagnosis," Jason said. "If you know a player is making his decisions here, start occluding it just before. And measuring the neural cognitive data every month, or whatever, gives you a way to track how precisely the occlusion training is working that you can't really get with response times."

"Are there players that can be helped?" Frank asked. "I don't have four hours every day to help every player the way I want to. What are the best ways to help each player?"

Jordan: "How do you identify those players?"

Frank: "That's what I'm saying—can this tool be used to identify those players?"

Jason and Jordan answered in unison. "Absolutely."

Jordan: "That's where we think we have an advantage over everyone else. We can pull that information out very easily."

"OK." Frank still appeared largely unmoved, his default position. "What makes your product more valuable is the neurocognitive side," he told them. "It's what makes it worth the price you want to charge." But, he added, "I've hit that point where I'm like, 'OK, I got to make this more usable.'" There are other companies whose products are no doubt sexier, he said. "But just because they're sexier doesn't mean they're going to help us perform," he said. "That's why I'm here." He is the experienced eye that can weed out the bullshit. Other members of the front office are just as likely to invest in the sexier company. "I have better buy-in and more people happy with that decision."

Jason, undeterred: "We know that the behavioral metrics are a pretty good way of measuring the accuracy and timing of the decisions," he said. "But it's the tip of the iceberg. The whole iceberg is underneath the water, measuring when those decisions are made."

Over the course of that summer, deCervo had collected two 40-minute sessions' worth of neural data from each player, plus 16 behavioral assessments that the players had done on their own throughout the summer, on the uHIT Mobile app or uHIT Virtual in the batting tunnel. They could say, for instance, that the data reflected a 39.7 percent decrease in chased pitches outside the strike zone or a 33.9 percent decrease in the time it took for them to click "Swing." They could say that the players were deciding on pitches 8.9 percent faster at the end of the summer than at the beginning, and they could show that that decision was happening, on average, four feet earlier in the full arc of the pitch. For each player, they offered a profile page with plots like this:

POSITIONS OF NEURAL SWING DECISIONS

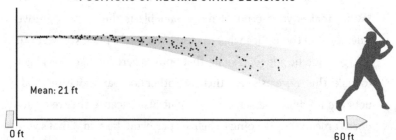

Scatterplot of neural swing decisions. Each dot represents the location of the ball in midflight when the batter decided to swing at it.

The dots correspond to every pitch that the player (in this case, it is Jason) responded to during his sessions. The plot is supposed to reflect the arc of the pitch as it travels from the pitcher's release point (left) to the batter (right). By clicking on any dot, you can see the precise distance from the release point that the player decided he wanted to swing. Every decision arrives at a different time. This could have to do with any number of factors: pitch speed, pitch type, release point, alertness, concentration or the arbitrary randomness of N. For each "pitcher" loaded into the game, though there is no animated windup, the release points and pitch trajectories still change, as in real life. With TrackMan data, which uses

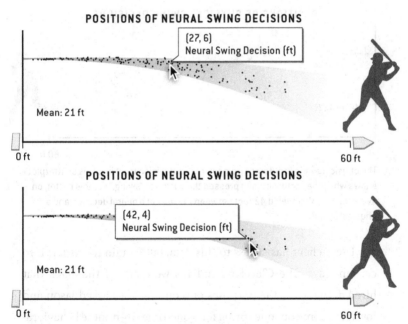

Users can go pitch-by-pitch to see when and where they made their decision along the life span of the pitch. At the top, the ball was 27 feet from the pitcher and 6 feet off the ground when the batter swung.

radar technology to measure baseball intricacies like spin rate and tilt, deCervo can allow teams to effectively practice against any pitcher in the game, including the ones who might be next on the schedule.

The "mean" in the lower left-hand corner shows the average distance when Jason decided to swing. This can be compared to another plot of where the ball is when he actually moved.

Another scatterplot would show where Jason made his decisions to "take" pitches. Of course, these cannot be compared with the physical swing plot, because he never actually swung. But the neural data still reflects the scatter of his decision-making.

POSITIONS OF PHYSICAL SWING DECISIONS

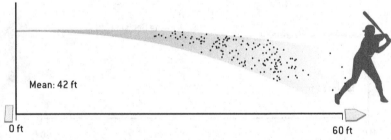

Mean: 42 ft

0 ft   60 ft

The physical swing is different from the neural decision to swing. This scatterplot shows when the batter actually pressed the button to "swing." For this batter, on average, the ball traveled 42 feet between the time of a neural decision and a physical swing.

Every client has access to this data, but certain teams use it in certain ways. The Cleveland Indians were one of the clubs that hired deCervo for the purposes of scouting. They asked Jason and Jordan to come out one spring for a one-time, 15-minute behavioral assessment of 57 players throughout various levels of the organization. Months later, when the season had ended, they sent deCervo

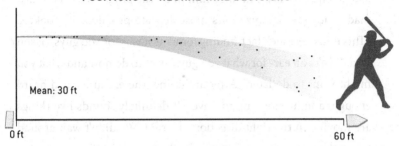

POSITIONS OF NEURAL TAKE DECISIONS

Mean: 30 ft

0 ft   60 ft

DeCervo can likewise chart exactly when a batter decided not to swing at a pitch.

advanced analytics recorded from the organization's TrackMan systems, which use radar technology to measure games on a pitch-by-pitch level. Armed with such pinpoint minutiae, Jason and Jordan plotted the uHIT results against the sabermetrics that once seemed overly pedantic: Z-contact ratios, out-of-zone takes, contact hits-in-play, balls seen. They excitedly wrote back things like: "O-Z Swing rate was anti-correlating with correct No-Go."

Though Frank had been critical at times in their meeting, Jason and Jordan left feeling sufficiently reassured about where they stood with the club. "That was pretty decent," Jordan said as they made their way to the parking lot. "We needed that," Jason said. At around 7:00 P.M., after Jordan had to leave town early, Jason picked up Frank and we drove down the road to the stadium, where the team was playing its penultimate regular season game. It was 101 degrees at first pitch, but the sun had slipped behind the stands in left field, and the sky took on the mauve tint of a ripening bruise. In Row P, directly behind home plate, the club's assistant general manager was chatting with a couple of scouts. Frank waved to him. We piled into the adjacent section. In the fourth inning, the assistant GM came over and took a seat next to Jason. For several minutes, he listened quietly as Jason filled him in on the progress deCervo had made with the organization's most delicate products: its rookies. "This is just as much fact-finding for us as it was for you guys," Jason said. "We're all ears for what you guys want to do now and what you might want to do later." After an inning, the executive had heard enough, and he stood up to leave. "It definitely sounds like things are moving in the right direction," he said. We didn't wait around to see the team win, securing themselves a spot in the postseason.

Despite the positive vibes from Arizona, a protracted silence ensued throughout the fall. Frank's big-league club finished among the worst teams in the National League. Average attendance at home games had fallen by more than 20 percent from the year before. It was looking like a long and reflective off-season. First, they watched the Chicago Cubs beat the Indians in seven games to win the World Series. Then they had to hear how strongly Chicago is positioned for repeated success. The saddest part: It was true. Kris Bryant (age 24) had won the 2016 Most Valuable Player award, while his teammate Anthony Rizzo (age 27) finished fourth. Three Cubs pitchers, including 26-year-old Kyle Hendricks, finished in the top 10 in the Cy Young race. The franchise was practically drowning in young, burgeoning talent. It did not seem like a coincidence that Theo Epstein, the Cubs' president, had helped assemble two of the finest championship teams in baseball history, vanquishing an 86-year title drought in Boston and a 108-year drought in Chicago. Epstein, a Yale graduate, was a Bill James acolyte whom the Red Sox hired, in 2002, when they could not lure Billy Beane from Oakland. He has been known to swim against the current. Several years ago,

while still with Boston, Epstein learned of a Cambridge, Massachusetts, company called NeuroScouting, which had created a video
game for baseball hitters that could supposedly measure things
like reaction times, dynamic hand-eye coordination and inhibitory
control. Epstein thought the game might be useful as a novel identification tool.

"You have to look toward other areas for competitive advantages," Epstein told David Axelrod on his podcast in January 2017.
The Red Sox tested prospects before the draft. "We could learn a
little bit about how their brain worked." In 2011, Epstein said, they
gambled on Mookie Betts, a spindly and virtually unknown outfielder from Tennessee, where many considered his best sport to be
bowling. Epstein liked him in part because of his predraft test
scores with NeuroScouting. Betts finished second in the American
League MVP voting in 2016.

Epstein's relationship with NeuroScouting, hushed for several
seasons, eventually got out. His acceptance of cognitive gaming as
a potential difference-maker was effectively an implicit invitation to
other teams that they should probably look into it, too. DeCervo,
born out of a laboratory at Columbia, as neophytes in business, did
not know about NeuroScouting when they presented their EEG/
fMRI paper in their first visit to the Sloan Conference, in 2013. By
sheer coincidence (they insist) they named their new company
NeuroScout. They needed to raise money, so they opened a Kickstarter page. A few weeks later, a cease-and-desist letter arrived in
the mail. Jason and Jordan claimed innocence, but agreed to change
their name. They lost the $8,422 they had raised on Kickstarter.

Though deCervo professed to be the only company that utilizes an actual neuroimaging technique to analyze players, it could still be hard to differentiate themselves to an audience that did not know much about neuroscience and likely did not really care, even if the science is purported to help them. I got a taste of this firsthand when I joined Jason and Jordan at the Baseball Winter Meetings, an annual détente involving the leaders of the front offices of all 30 Major League teams. It is a traditionally feverish week in a sumptuous tropical setting. That year, to my chagrin, it was in Oxon Hill, Maryland. The blustery winds outside matched only the blustery prognostications inside. In December, every team can claim to feel pretty good about their upcoming chances. Thousands descended on the Gaylord National Resort and Convention Center—client seekers, job seekers, internship seekers and autograph seekers, along with assorted consultants, licensers, marketers, media, agents and a smattering of actual executives. The GMs mostly holed up in hotel rooms, engaging in trade conversations among themselves. The media milled about the hotel lobby. Downstairs, the convention center hosted a trade show and job fair. Those attendees were actually given a lanyard with "JOB SEEKER" in bright blue lettering on the label, so they could be distinguished (for better or worse) as they wandered about the lobby. It can be a humbling experience. "Could you imagine?" Jason said. He was relieved deCervo did not need that lanyard. For weeks prior, they had been furiously messaging team contacts to set up face-to-face meetings, and they managed to get a full day booked.

They arrived at 7:30 A.M. and quickly apprehended a cocktail table along the balustrade of a second-story patio deck within a 19-story indoor atrium, which contained a murmuring koi pond, a clapboard house with green shutters and an inexplicable chimney, and a 65-foot-tall Christmas tree made from synthetic glass. The rising sun was just illuminating the steepled skyline of Alexandria, Virginia, as seen through a 20-story wall of windows. Jason and Jordan looked quietly at their laptops. They had prepared 15-minute presentations. Shortly after eight, six men arrived in their off-season uniforms (blue jeans, button-downs and blazers), pulled up chairs and fired off questions. How does the neural work? How do we set this up? How do we prove it works for scouting? Do you have controls?

And so the day began. The patio assumed that just-before-the-final-exam buzz of whispered conversations among competitors eyeing one another warily while standing in broad daylight. It was necessary to be seen but inaudible. Team executives hopped from table to table like eager speed daters. One general manager arrived at deCervo's table flanked by three subordinates. The subordinates never said a word. "Guys, why don't you ask a question?" the GM said at one point. "Don't worry about it, they're shocked that I'm actually here." As were Jason and Jordan. They took it as a good sign. "We know sports tech," Jason said. "We know you're trying to figure out what works. We've been doing this for the last two years. We've designed uHIT to fit neatly into your operations, but at the same time incorporate cutting-edge neuroscience."

They had recently visited with a vision training company, another GM said. "How do I know that your shtick is different from anybody else's shtick?" he asked.

"Vision training is great," Jordan said. "But the thing is that there are a disproportionate amount of players in the Major Leagues that have 20/20 vision. And not everybody is an All-Star. Something else is going on here. There's object tracking, motion tracking, all these other things that have to combine with the motor portion of the brain. Then they have to decide to swing or not swing."

The GM nodded. "That makes sense."

"We've had several studies published on our work," Jason said.

"Peer review is nice and all," the GM said, "but what does that mean for us?"

Suddenly, there was a high-pitched squeal from somewhere in the lobby, stopping everyone in their tracks. An alarm had been triggered. Jason ran off to see what was causing it, if it could be stopped. But the conversation at the cocktail table ceased, succumbing to the strident siren, which bleated on for several minutes. "It would've been better if dog shit had rained from the heavens," Jason would say. The GM and his associates thanked them, shook hands and left.

The alarm did not completely dampen their enthusiasm. In a quiet moment, Jason leaned in to Jordan. "I think it's an easier sell the way we've broken it down." He was referring to the uHIT Mobile, Virtual and Neural platforms. Jordan agreed. "Definitely." But each team that sat down seemed to share consistent concerns. "Your organization is like a battleship," Jason explains to one group. "We know one product isn't going to turn the ship at once. We made it more piecemeal, so you can try something out at an affiliate level over the course of a season. If it works at one level, you can explain it at all levels." Jordan tried to read the expressions of those he was sitting across from, evaluating their poker faces. "Teams don't want to tell us much," he told me later. Jason repeated his iceberg analogy. "That gets through."

The eighth meeting was late in the evening. It was with Frank. He was the only member of the organization to show up. For several minutes, he apologized—the team had not yet put together its 2017 budget, and therefore could not say for certain whether or not their relationship with deCervo could continue. It was disappointing news.

They trudged back to the hotel room they shared at a Crowne Plaza in Alexandria. Exhausted, Jordan collapsed on the bed. The day had felt like a college job fair. But they were confident they had earned some trust on the other side of the table. They felt like they at least belonged there. "Last year, I was sneaking into things, e-mailing people," Jason said. He was hawking business cards in the lobby. Jordan had told him that is not how he wanted to do things this year. "Jordan wants to plan things out a lot more, which is good, it's very helpful. I think it went a long way." They also had more complete data to show, a product menu, a variety of platforms. "It wasn't until the end of this season that we were able to convince ourselves that we had something to sell to somebody that we didn't feel sleazy about."

Yet a sense of security, as a startup in the sports industry, can be fickle, if not tragicomic. I looked on the next morning as Jason and Jordan again set up their laptops at the cocktail table in the atrium, waiting for a designated team representative to arrive for his appointment. It was the final day of the Winter Meetings; all around them people were rolling suitcases through the lobby and shaking hands good-bye. Black SUVs queued around the hotel's cul-de-sac. Jason and Jordan waited for an hour. The guy never showed.

After more than two hard-fought years, deCervo had established relationships with almost every team in Major League Baseball. They had collected a full season's worth of data from two franchises, and signed contracts with another three, not including a professional team in Korea. They expected that at least three organizations would be utilizing their products throughout the upcoming Spring Training, with the hope that a longer-term connection might follow. Only two years earlier, Jason had presented one of his early EEG and fMRI papers at Sloan to the snickers of some Major League scouts, who were exhibiting their new approaches to advanced analytics. "They were talking about regression models and that kind of stuff they were doing, in the 'Moneyball' vein," Jason said. He had asked them if they were curious about digging down even further, to the cognitive level, to extract what exactly it was that might be generating that wins-above-replacement number or strikeout-to-walk ratio. "Most guys looked at me like I was crazy. They were like, 'Nah, we don't need to drill down to that level,'" he said. "I'm like, OK. I'll see you in 10 years."

But the company's viability remained uncertain. They had not yet seemed to convince teams that the pathway to a competitive advantage in baseball was through the brain. It was even fair to wonder how far they were drifting from the empiricism that set them apart. Neuromonitoring still distinguished them, but their correspondences at the Winter Meetings seemed to be pushing them toward the video game model. The marketplace was starting to dictate the product offering.

"Teams will often dictate what they want," one National League performance director told me. "They made an assessment tool that I think is scientifically valid, but there are caveats. They've had to gamify a lot of aspects, to appeal to a new generation, to players." The cogency of the data deCervo was offering would not matter as much as a usable, appealing framework.

It was hard not to be reminded of the comments that John Krakauer gave at the end of the Santa Fe workshop. He can appreciate the tacit wisdom of professional coaches and athletes, who have never needed to *know* that engagement of their fusiform gyrus is what allows them to swing at baseballs at the right time. That knowledge has never been necessary in order to be able to do it. Now that science is trying to intrude on their space, there is pushback. Sports have thrived largely ignorant of academic analysis. Hockey players smoked in the locker room at halftime and baseball players scoffed at lifting weights. Some of that, unquestionably, has changed, but a yawning gulf remains: "How do we reconcile all the buried knowledge of sport science and the boring work being done in laboratories?" Krakauer asked.

There are doubts that coaches know the most efficient practice methods, just as there are doubts that motor laboratories can produce useful breakthroughs while conducting tedious experiments on mostly white male undergraduates. "This is not working," said Xavi Schelling, the director of performance for the San Antonio Spurs. "We are wasting our time, and we are losing the credibility of the athletes. And they are right. We have to refigure out the whole model and start asking the right questions." Vincent Walsh, the neuroscientist from UCL who works with professional soccer teams, mostly incorporating behavioral psychology, said of such disconnect between teams and science: "I haven't gotten a flavor of what you want from me. There is a distance between the lab and real world."

Krakauer was not quite ready to pin all the blame on the scientists. "Everyone has a notion of the superstar, and everyone thinks they know who the best trainers and the best coaches are," he said. "Are we ever going to go beyond that gestalt feeling to something that is actually defined by science?"

A few extraordinary performers of late have been generous with their brains. A study done recently at the Medical University of South Carolina brought in Alex Honnold, the world-class free-solo climber. Inside an fMRI, they showed him 200 images of alternately disturbing, disgusting or invigorating scenes. The researchers wanted to look at Honnold's amygdala, the brain structure that is implicated in recognizing and responding to fear. They first wanted to see if he had one (he did). Then they looked at how it functioned compared to other people's. Honnold is someone who blissfully scales steep cliffs without a rope. In Yiddish, he would be called a *meshugener*. In Yosemite, he is called No Big Deal, for the nonchalance with which he approaches his climbs. The amygdala is responsible for producing the changes in heart rate, breathing and pupil dilation when our brain senses danger. The images shown to Honnold were designed to be graphic enough to send his fear-monitoring center into overdrive. But the researchers found that Honnold's amygdala remained inert. It was as if he had been watching Golf Channel. In fact, they wondered if Honnold had fallen asleep (he had not).

Another study recruited Neymar, the sensational Brazilian soccer player, to a scanner in Osaka, Japan. In the bore, he was asked to rotate just his right ankle leftward and rightward in synchronization with a series of metronomic sounds. What they found was definitively *less* cortical activation in Neymar's foot motor regions than in even those of other professional soccer players participating in the study. The researchers accepted this as a sign of efficiency: Neymar's skillfulness with his feet meant that he could conserve neural resources "probably with higher reproducibility and less effort." They took the narrative a step further by also volunteering to "assume that this fundamental capability of his football brain could allow him to spend neural resources to focus more on cognitive aspects during a football game, such as anticipating/predicting and detecting the actions of other players."

Krakauer might argue that these studies did not truly advance the understanding of talent any more than noting how much a linebacker needs to be able to bench press. His contention has been that the scientists need to understand what it means to be skilled before we should be popping superstars into scanners and drawing too many conclusions about amygdalae and neural efficiency. "We all can watch the best dancer on the stage and you can have people vote, 'Yup, that's a great dancer,'" he said. "But what are they talking about? What have they detected? And it's actually hard to know what that is."

DeCervo was trying to use its insights about rapid perceptual decision-making to help baseball teams win baseball games, but, to their frustration, those teams were increasingly concerned about its footprint on the players' already heavily loaded schedules. And players were concerned about being guinea pigs for testing. Some feel the information could only be used to hurt them in the future. And why spend 40 minutes with an EEG cap on your head if there is no immediate performance benefit associated? And that is at the rookie level. At the higher levels, time is considered even more valuable. Is it worth pulling a top prospect away from batting practice for 40 minutes? Or rousing him early on a Saturday morning? "It's definitely a lot," a member of a National League analytics department told me. "The value has to be significant to commit to that sort of thing." Teams are currently weighing that value. He added, however, "If their claims can be substantiated, then honestly there is something to it."

Anyway, there is also a darker concern. Baseball injuries and ailments have long been treated as the team's issue, and players are routinely given medical evaluations before they are accepted as part of any trade deal. But, in the case of medical information that is not so readily available, like EEG recordings, the situation gets murkier. Baseball has already found itself in hot water in recent years for conducting DNA tests on prospects from Latin American countries to properly determine their ages. But in a few years, this could begin to seem trite. At Sloan, in fact, one of the sponsors was a company called Orig3n, promoting DNA tests at the swipe of the cheek. They had a booth in the main hallway with cotton swabs by the bucketful. One National League coach outlined the potential conflicts that could arise from the proliferation of more of this kind of testing. "If I do a whole genome isolation and I find this guy has [the mutation in one of the genes associated with] early-onset Alzheimer's, which we know can present as early as 25, what do you do? If he's graded as a Round 1 draft pick, do you take him and trade him?" Some teams, he implied, probably would.*

In 2016, the San Diego Padres' general manager, A. J. Preller,

---

* In 2005, the Chicago Bulls asked that forward Eddy Curry take a genetic test to determine if he had a rare mutation affecting his heart. Curry refused and wound up signing with the New York Knicks, who never tested him. This was before the Genetic Information Nondiscrimination Act of 2008, which prohibits employers from using information from genetic tests for certain hiring practices. It has yet to be tested judicially over the practice of drafting or trading baseball players.

was suspended for 30 days without pay after an investigation into a trade of pitcher Drew Pomeranz to the Boston Red Sox. The league found that the Padres had not been up-front about Pomeranz's medical history, going as far as instructing medical trainers to maintain a file of medical information that they intended to share publicly and another they intended to hide. All Major League teams are expected to feed players' medical information into a database known as the Sutton Athlete Health Management System, and when a trade is made, passcodes are swapped, and teams can access that information. According to Article XIII(G) in the recent Collective Bargaining Agreement, it is limited to medical and health information as assessed by the club trainer or physician. But at this point, neural markers are not considered to be medical information. Instead, in July 2017, the MLB Players Association released a three-page appendix to the CBA stipulating that information gathered from wearable technology—including off-field devices for measuring performance—can remain confidential to the team and available to the player upon request. Teams have told me they consider brain data of the players to be proprietary, even though, technically, they are not the ones collecting it. DeCervo is, and they are permitted to keep the data, too.

DeCervo and NeuroScouting are not alone in this nascent and ambiguous neurotech space. For one, there is Halo. But I also visited another cogni-training company, NeuroTracker, for a story in *The New York Times* in January 2017. Its creator, Jocelyn Faubert, of the University of Montreal, has been attempting to study perception and action coupling by using the concept of multiple-object tracking, an experimental device designed by the psychologists Zenon W. Pylyshyn and Ron W. Storm in the 1980s. In Faubert's iteration, eight spheres bounce randomly and rapidly around a three-dimensional cube, and you are assigned to track four of them with your eyes. After about eight seconds, the bouncing stops and you are asked to recall which balls you were tracking. It is like a three-card shuffle. What Faubert discovered, almost purely by chance, is that athletes were much better at the experiment than nonathletes. They could also improve at a much faster rate. Faubert created a game: If you are able to successfully remember which balls you were assigned, you move up a level and the balls speed up. If not, you move down and the balls slow down. From this, NeuroTracker the game was born. Today, it is in more than 550 elite training facilities around the world, including ones used by Olympians, NHL teams and European soccer league teams. Matt Ryan, the quarterback for the Atlanta Falcons, who played in Super Bowl LI, told me he uses NeuroTracker three times a week during the season.*

Faubert, an energetic and ebullient Montreal native, said it did not surprise him that expert athletes revealed differences in how they are able to perceive and react to external stimuli. "Their brains are special," Faubert told me. "I got a lot of reaction from that. Because people were like, 'What are you telling me? That guy, who doesn't have a degree, never went to university, can barely read—you're saying that guy is smarter than me?' I say, 'Well, that guy can use his brain much more efficiently than you, for certain things.' It's a form of intelligence. You don't have to be smart to do physical activity. But you have to be smart to become the best in the world at it."

---

* His opponent in the Super Bowl, Tom Brady, also endorses "cognitive training" through a company called Posit Science. The 29 "TB12 BrainHQ" exercises include ones designed for "people skills" and "intelligence," and a ball-tracking game similar to NeuroTracker.

It also seemed reasonable that some players would believe a game like NeuroTracker could be helpful. Athletes would certainly seem to be relying on attributes like spatial awareness, visual tracking, concentration and working memory all the time when they play, and these were the attributes that NeuroTracker purported to enhance. Sports such as football, basketball, hockey and soccer involve dynamic, shifting, complex scenes of action unfolding all around, a blizzard of stimuli pouring onto the sensory system. Maintaining order amid the chaos—not letting your eyes be overwhelmed by the dummy spheres—is part of what can lead to success.

But clinically establishing that high scores in NeuroTracker can translate into high marks on whatever field you play on has been more elusive. Around the turn of the century, Edward Thorndike and Robert S. Woodworth each concluded that practice at one skill—memorizing poetry, for example—could not help a young student with her scores in mathematics, unless there was something elementally binding about the two tasks. The controversy of whether skills can "transfer" has persisted since.

Some researchers such as A. Mark Williams, the chairman of the Department of Health, Kinesiology, and Recreation at the University of Utah, challenged the notion that tracking bouncing objects in a simulation could train or quantify anything other than a person's ability to track bouncing objects in a simulation. "I've never seen a soccer player chasing multicolor balloons around on the field," Williams told me for the *Times*. "It's just not what soccer players do." What soccer players do, he added, is read patterns of play, anticipate what might happen next based on movements of teammates and opponents, and identify familiar sequences as they unfold. This is the "inside" knowledge, built up over time, that promotes the effectiveness and efficiency that Anders Ericsson argues are the hallmarks of expertise. To actually get the benefits of practice, you need the right sporting context so that the brain can prepare for what it might see in a game. Randomly bouncing balls, a neuroscientist from Arizona State, Rob Gray, told me, is so contrary to the real game of basketball as to be essentially meaningless. "The whole point of watching a basketball scene if you're a point guard is that it is structured," Gray says. "Picking up that structure that's specific to your sport is highly important."

DeCervo believes it has avoided making the "transfer" mistake in designing uHIT. Their simulation is at least authentic to the setting of baseball, on any clement and windless summer day. They added a "batter's-eye" view, giving hitters a more plausible perspective of the angle of the incoming vector. Previously, it had been more of an umpire's view. They are advancing toward the live-action video clips that Jordan showed me. Pierre Beauchamp, a well-regarded sports psychologist based in Montreal, has spoken with deCervo about using live video to train hockey players. "You can't use avatars," Beauchamp said. "Avatars don't work. There are no visual cues. The research has proven that." Others have advised them to keep the video game feel, but add a pitcher's likeness. Still others think they have done enough. The National League performance coordinator told me he would hardly change a thing about deCervo's platform. "Do I think that deCervo or EEG like that can and will be utilized in the future?" he said. "Yes, I do."

Their five-minute presentation at the Sloan conference's startup competition would have given any onlooker the same impression. When Jason mentioned that they had recorded more than 200,000 pitches' worth of data from professional players, totaling more than

35 million individual brain responses, the judges' eyes lit up. Ultimately, though, deCervo lost in the seed-stage category to a data company from Seattle aimed at streamlining football coaches' stat sheets. Jason's grass perfume only carried them so far.

He intended to mill around the convention center to mingle and network and give away yet another deCervo hat. To Jason, the world can seem an endless procession of hands to shake. He has been described to me, by a scientist, as a "good salesman" and, by his mother, as a "real pain in the ass." But Jordan had long since had enough. It was nearing 9:00 P.M. We walked out into the frigid night and found a pub, where the Cleveland Cavaliers were playing the Atlanta Hawks. For a while, we sat in silence, watching the game.

"If we don't succeed," Jordan said at one point, "I still think the idea will succeed."

6.

SEARCHING FOR
THE MOTOR ENGRAM

THE INTELLIGENCE IN OUR SKIN

We marvel at the artistry of action. To move smoothly, to glide effortlessly, to leap and prance, on the field or on-stage, is to epitomize the wondrous but not limitless kinesis of our human form. When LeBron James precisely delivers a behind-the-back pass, or Mike Trout races down a fly ball, we rarely stop to consider the sheer magnitude of the computational demand. We can appreciate the aesthetic, and that is hardwired in all of us. It requires only seven small lightbulbs attached to the major joints for us to distinguish the form of another human in a darkened room from that of an ape, a bear, a bookshelf, a robot. This *only* works, however, if there is movement. In motion, the lights form the shape of something fundamentally recognizable to us, something distinctly human. When researchers ventured into the jungles of Brazil, they found the ease with which the Mundurucú, an indigenous Amazonian tribe, detected the characteristics of a human's movement using simply points of light attached to an otherwise dark body to be in accordance with laboratory participants across three decades of research. Our recognition of biological motion is universal. Our captivation with it has been evident since Neolithic cave paintings of swimming and hunting. Our understanding of it, in turn, has been vexed.

Needless to say, action is not simple. It does not spill forth like paint from a tipped can, though it can sometimes appear just as smooth. It is the sum of many dozens of small parts, systematically arranged into the mosaic we have come to instinctively recognize. But the pattern and organization of this arrangement have been hard nuts to crack. Many years after Sherrington introduced the world to the significance of reflexes, a fundamental question about motor skills remained. A tenable theory for serial movement—the complex strings of motion involving thought and planning that generally form the basis of what we call a "skill"—remained elusive. When a pianist sits for a performance, what enables his fingers to consistently glide across the keyboard and strike the correct keys? Or when the alpine ski racer Lindsey Vonn plunges headlong down a course, how does she guide her body to maneuver within the poles? She needs to maintain a map of the course in her head, and also the precisely timed and coordinated motions to achieve that task. But are they memorized or improvised? It is a lingering question that has at least one haunting historical antecedent. In 1401, the German pirate Klaus Störtebeker was captured and, just before being executed, struck a deal so that each man he walked by after his execution would be spared. As legend would have it, his headless body managed to walk past 11 of his men before the executioner tripped him.

Clearly, Störtebeker wasn't improvising. So was the movement then "memorized"? Was it what we call "muscle memory"? According to Sherrington, movement was based on the premise of simple reflex action and inhibition. These stereotypic movements were the basic units for action. Complex sequences could be formed by the combination of coordinated reflexes, like a Rube Goldberg contraption. Every action triggered something that comes next, incorporating feedback from the action before it. He seemed to recognize that he might be oversimplifying things, but he simply left it at that. "To subsume the whole of human behaviour under what has been called 'reflexology' might further be taken to mean that the roof-organ of the brain reacts simply on the reflex plan," Sherrington said in 1937. "But that there is much which contradicts. Our inference has to be that we are partly reflex and partly not." Few questioned Sherrington's view. In fact, his ideas helped spawn a new camp of psychologists, the Behaviorists. The early adherents—John B. Watson, Edward Thorndike, and B. F. Skinner in the United States; Ivan Pavlov abroad—were not remotely interested in any role of conscious thought in action, arguing that it was too intractable to study. They focused solely on what was observable in behavior. Sequences of movement were perceived as those steady chains of reflexes—the domino effect of elements exciting elements, one after another, in a serial succession. The whole became the sum of its parts. Information from the muscles traveled up to the brain, triggering the next response, on a loop. Neurons largely lay dormant until prompted by an impulse. When the required action was completed, the system settled back into a latent state. Long sequences of action—from orating a speech to sawing a log—could be explained without the need for any evidence of the mind. "There may be a few scattered ideas possessed by the higher animals," Thorndike wrote in 1901, "but the common form of intelligence with them, their habitual method of learning, is not by the acquisition of ideas, but by the selection of impulses."

Another camp, the Structuralists, took up an opposing view. They were not so ready to divorce the relationship between consciousness and action. Rather they suggested that actions were borne from intentions, which are formed upon an aspect of mental imagery. Edward Titchener, in describing how this imagery helped his mouth to form the words he intended to speak, wrote, "I hear my own voice speaking just ahead of me." It is the lazy man's lecturing dream. Consciousness would hardly be of more use than a teleprompter. It would take some time, but eventually psychologists began to yearn for more empiricism from both sides. "A behaviorist colleague once remarked to me that he had reached a stage where he could arise before an audience, turn his mouth loose, and go to sleep," wrote Karl Lashley in 1951.

Lashley was a student of Watson's, raised in the Behaviorist cocoon. A weedy man with tousled hair, he was born in rural West Virginia and, as a boy, grew fascinated by his mother's sewing machine. Later he took a keen interest in carpentry. He was possessed by the desire to understand the mechanics of things and often felt unsatisfied with what explanations there were. This included his feelings about the human brain. His iconoclasm often put him at odds with the rest of the neurophysiology field at the time. "He was especially adept at marshaling convincing evidence which demolished every existing theory of learning, including his own," wrote Frank A. Beach in his biographical memoir. Interested in understanding motor skills, he once calculated the speed of the finger movements by a pianist performing a cadenza and compared this to what the known speed was for neural transmission. The discrepancy was enough to get him thinking about the Behaviorists' explanation of the reflex chain, in which inputs triggered the next in a series of movements. How could a sensory message be sent up to the brain, processed and then sent back as a motor impulse for each finger press? The numbers did not seem to line up.

Furthermore, once the initial impulse was sent to initiate the chain of command, what exactly dictated how the resulting actions would follow? How was the order assigned? Lashley was troubled by a common problem he would experience while typing. He often made errors, either in misspelling or the doubling of a letter, such as typing *t-h-s-e-s* instead of "these" or *l-o-k-k* instead of "look." If the mechanism for activation was so structured—if the loop was truly as mechanized as it was thought—he should not be consistently having these problems. And what of a person whose feedback is interrupted, like a hemiplegic? Is he wholly incapable of producing a pattern of movement?

Lashley's own research on localization prompted him to think more critically about the organization of the brain. He did not buy into reflexology. Instead, he searched in vain for what was known as the "motor engram"—a place in the brain where learned movements were thought to be stored in memory, like a locker. Indeed, if emotional memories were stored in certain locations, it would follow that muscle memories—those procedural representations of an action—are likewise contained in discoverable cubbies as well. But, Lashley wondered, if such a locker could be accessed, how would the motor memory unfold? Would it trickle out in pieces or does it actually have the capacity to store the entire link? Are we to presume that, if one area of the brain is stimulated electrically, out will pop an entire piano concerto?

In 1951, Lashley composed his quandaries in a paper he called "The Problem of Serial Order in Behavior." In it, he said he viewed the brain like the surface of a lake. "The prevailing breeze carries small ripples in its direction . . . Varying gusts set up crossing systems of waves, which do not destroy the first ripples, but modify their form." A tumbling log sends out more ripples, a speeding boat more still. All of these influences, simultaneously and interchangeably, alter but do not disfigure the lake's surface. Nor is the lake's surface ever static. He pointed to both the rhythmic system—the principle that guides the unconscious movements of our internal organs—and spatial orientation as evidence of a more elaborate nervous system than the Behaviorists would care to admit. The brain, it seemed, was capable of integrating a wide array of elements, constantly. Lashley wrote simply that his article was intended to deal with "the logical and orderly arrangement of thought and action." He had dismantled the narrow formulation of feedback triggers and replaced it with the view of a more flexible, adaptable brain, which orchestrates behavior with something the Behaviorists were reluctant to consider: a plan.

Jörn Diedrichsen, a researcher at the University of Western Ontario, about two hours west of Toronto, was flipping through the pages of Lashley's work for the thousandth time recently with astonishment. "It's sobering," he said. "I find his writing so luminous and clear and relevant today. I read those papers and say, 'Not much has changed.'" I met Diedrichsen one winter morning in his office on the second floor of the Brain and Mind Institute, an arrayed laboratory within one of several neo-Gothic structures at the center of a wide and neatly appointed campus. We walked down to the cafeteria and grabbed lunch, which for Diedrichsen consisted of a 20-ounce Diet Coke and a McIntosh apple. He devoured the apple, core and all; the soda he sipped. He is about five foot ten, with a cyclist's frame, buzzed hair and a wide mouth that smiles often. Born and raised in northern Germany, near the coast of the Baltic Sea, he thought he would grow up to be a physicist and has retained that computational interest throughout his life. But when Germany required he fulfill a social service or join the military, he chose the former. He was sent to northern Norway, above the Arctic Circle, where the days were short, or sometimes never came at all. There was a psychiatric ward; that became his social service. He entertained and assisted the patients, some of whom suffered from schizophrenia, and he became struck by the complexity and fragility of the brain. So he decided to study psychology at UC Berkeley, where he met Professor Rich Ivry, who studies motor control and neurology, and they clicked. He knew what he wanted to do. He wanted to find the motor engram, and he chose where he wanted to look: our hands.

Diedrichsen considers our hands to be the Swiss Army knives of the body, capable of practically whatever we need whenever we need it. There seems no end to their utility and complexity. Your fingers can pluck guitar strings and tickle the ivories. They can grip a rock face or perform microsurgery. "It is in the human hand that we have the consummation of all perfection as an instrument," Charles Bell, the early nineteenth-century physiologist, once wrote. In the 1880s, John Hughlings Jackson spoke of "evolution and dissolution," that the most highly evolved functions of the motor system were also the ones most susceptible to acute disease. Years later, Roger Lemon, a renowned neurologist at University College London, has suggested that there could indeed be a link between amyotrophic lateral sclerosis and the most highly evolved system we as humans possess, our hands. "Tsarina of motor abilities," the Russian movement scientist Nikolai Bernstein once said of dexterity.

Darwin is generally credited with being the first to consider the potential impact of our evolution out of the trees and into an upright walking posture, which freed our hands for other tasks beyond swinging from branches. "Man could not have attained his present dominant position in the world without the use of his hands, which are so admirably adapted to act in obedience to his will," Darwin wrote in *The Descent of Man*. That will, more than likely, involved our unique use of tools. Though early *Homo sapiens* were more than likely not performing surgical incisions, the adaptation of long thumbs and robust finger bones in humans helped, for instance, in the skinning and carving of a hunted carcass. Neurologist Frank R. Wilson, in his book *The Hand*, echoed a popular view that the development of human hands could have coincided with the enlargement and enhancement of the human brain beyond that of other species. Because, as he wrote, "there were so many new places that hands could *be*." In that transition, the hand became more than just a utility; it explored and discovered. It became "a divider, a joiner, an enumerator, dissector, and an assembler." It could also be "loving, aggressive, or playful. Eventually, it found in the intimate touch of grooming the secret to the power of healing. It may also have been the instigator of human language."

The hands of all primates *look* similar, but their functions vary widely. Some can only grip tree limbs; others can handle tools. It was Oxford's Charles Phillips, half a century ago, who first discovered that primate dexterity might have something to do with the strength of the connections between the spinal cord and the cortex, a concept now widely accepted. "The capacity to execute relatively independent finger movements," Hans Kuypers wrote in 1981, "is characteristically provided by the corticospinal pathway." Our evolutionarily honed spinal pathway enabled us to manipulate tools, and by doing so, we enhanced the potentiality of our cortices, expanding the reach of our hands ever further. And, somewhere along the line, we learned how to grip a fastball. The neurophysiologist William Calvin once suggested that the arms of female hunters, before agriculture, were relied upon for killing their prey with the throw of a stone. Because the women also had to nurse the children, they developed a trick to keep the baby quiet. Babies sleep better if their heads are resting up against a beating heart, so the mother would hold the child in her left arm, freeing her right arm for the hunt. This legacy still lives in us, Calvin wrote, which is why around 90 percent of us today are right-handed.

The origin of the handedness tale is still being debated, even though there is evidence of hand preference among hominids going back two million years. A study of the wear patterns of 32,000-year-old Neanderthal teeth suggested that 88 percent were right-handed. A survey of artwork dating back 5,000 years showed the right hand was used 93 percent of the time. Babies almost always suck the thumb on their right hand in the womb. Those who don't are more likely to grow up left-handed. The arm you throw with is seldom uncorrelated with the foot you kick with. Blonds may be more likely to be left-handed than non-blonds. Mice are almost equally likely to be born a lefty or a righty, but most parrots tend to pick up food with their left foot, the same as male cats, only with their left front paw. Female cats are said to prefer their right paw, like most humans. Most people thought monkeys had no hand preference, until a study found that orangutans seemed to fancy going with their left. A long-standing myth that left-handedness was related to devil worship possibly came about because the Latin word *sinistra* meant "left." Al Hendrix believed that, and he used to scold his left-handed son, Jimi, who eventually learned to play guitar with both hands. But if you were born right-handed, you almost certainly write with the same hand. The impact of handedness can even promote physical changes in the brain. Studies have shown that the sulcus, or groove, that lies along the posterior edge of the primary motor cortex is deeper on the side that correlates to your dominant hand. The development of gray matter is greater—it bulges out more—making the sulcus deeper.

Why we can feel so comfortable in this lateralized formulation of our bodies most likely has something to do with the lateralized formulation of our brains. The hemispheres in our brains, though they look like mirror images, also appear designed to operate toward different objectives. When this hemispheric divorce was discovered, midway through the nineteenth century, it was thought to be a consequence of poor design. A perception arose of the sides being unequal, which fomented the opinion that the left hemisphere served as the "civilized" guardian of the more "primitive" right hemisphere, which could act out if somehow unleashed. The debate might have captivated the imagination of at least one author, Robert Louis Stevenson, who in 1886 published a novel about a man's inability to harness the animalistic tendencies of his right hemisphere in *The Strange Case of Dr. Jekyll and Mr. Hyde*.

However fictionalized the issue became, certain notions about the lateralization of the brain and its links to behavior persisted until the early twentieth century. In 1903, the Ambidextral Culture Society in London was formed as part of a wave of interest in promoting dual-handedness as a method of sparking improved cognition. "Our desire," John Hughlings Jackson said, in remarks he made in a lecture titled "The Advantages of Ambidexterity," in 1904, "is that teachers, parents, and those interested or occupied in the work of education may join our ranks and become active agents in advancing the cause of bimanual training." Members would learn, for example, to play the piano with one hand while writing a letter with the other. Eventually, they grew tired of both.

This idea about lateralization has continued to circulate. Psychological studies on split-brain patients in the 1960s reinvigorated the impression that the hemispheres had varying strengths and weaknesses. A "right-brained" person became one who is better in touch with her emotions, while a "left-brained" person is better at deduction and problem solving. Most of us have heard, at some point, that creativity derives from our right brain and focus and rationality from our left. Neither has stood up to significant neuroscientific scrutiny.* When it comes to motor control, specifically, scientists have long since begun to recognize that lateralization was not about competition—developing the left side of the brain (considered to be responsible for the ability in most people to produce

---

* At the very least, this persistent myth seems to disregard the corpus callosum, the thick bundle of fibers bridging the hemispheres like a spider's web of connections.

and process speech) does not come at the expense of our right. Nor is one side solely responsible for controlling the other. The two sides work cooperatively, and they need to. Manipulating the hands to perform the complex tasks we do every day requires a team effort.

By the early twentieth century, there was still no consensus for exactly how the human cortex was organized to fulfill such exacting behavioral requirements. The prevailing belief, since Hitzig and Fritsch, was that the motor cortex was somatotopically arranged: Different regions corresponded with the control of different body areas, like an internal map. In 1940, Edgar Adrian discovered that the size of the cortical representation for a given body part in animals depended on how important that part is for the species. For example, monkeys had hands and lips that occupied almost their entire somatosensory cortex, the receptive area for touch. The nose region of the Shetland pony occupied a cortical territory as large as that for all its other body parts combined. "If we could apply electrodes to the exposed cortex," he wrote in 1943, "there is no reason why we should not map out all the receiving areas in man as we can in animals."

Indeed, a few years later, a Canadian researcher named Wilder Penfield—another former student of Sherrington's—managed to find more than just man's sensory areas. He published a map of locations along the cortex where stimulation produced differing movements of the human hand. The map, once fleshed out to include the greater body parts, became known as the "human motor homunculus," derived from the Latin word for "little man." In a 1950 book with Theodore Rasmussen, *The Cerebral Cortex of Man*, a garish image appeared of the brain and the specific areas that are

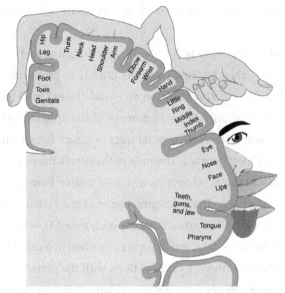

The human motor homunculus, Penfield's orderly
representation of the motor cortex.

in control of specific body parts. In this view, the motor cortex is a
map of muscles.

The hand and fingers accounted for an extraordinary expanse
of cortical real estate. The illustration helped to conceptualize Penfield's earlier lectures, although there is some disagreement about
whether it caused more harm than good. For neuroscience, it became a marketing hit—a memorable and easily interpretable image
for students in science classes nationwide, which still survives in
many textbooks. But even Penfield knew the illustration was exag-
gerated and misleading. His map "cannot give an accurate indication of the specific joints in which movement takes place," Penfield
wrote in 1950, because movement typically occurs at multiple joints
simultaneously. Penfield's map was not nearly detailed enough to

go beyond a vague estimation, which was insufficient for complex movements.

What is also missing from Penfield's diagram is any sense of interaction with other brain regions critical to producing movement. The motor cortex is not on an island distinct from its counterparts. In fact, only 30 to 40 percent of the approximately one million axons in the corticospinal tract originate from neurons in the primary motor cortex. The map of the cortex needs context. It is more like a subway schematic, with information flowing in reciprocating directions. The key regions serve as the various boroughs of New York City. The primary motor cortex (or M1, which is how we now refer to the principal brain area involved in motor function) is insignificant without its connections with the premotor cortex, supplementary motor area (SMA), primary somatosensory cortex and posterior parietal cortex. The premotor and supplementary cortices boast their own codependent connections, running routes to the brain stem and spinal cord to produce movements sometimes without the need for M1 at all. Baseball hitters, whose SMAs are inhibiting their swings, are quite possibly relying on the corticospinal connection between that region and the musculature itself. Projections to the basal ganglia are so precise it is said to contain its own motor homunculus, which might seem redundant, but for its critical role in action selection, preparation, execution, and sequencing. More than half the neurons in the brain are located in the cerebellum. An intended movement aligns with the actual movement on the strength of many corrective cerebellar signals, like a friend who adjusts your tie before heading out the door.

Penfield, like his mentor Sherrington and his description of reflex action and the production of complex actions, had left out critical neural architecture. It is why science needed someone like Karl Lashley to think more holistically about the building blocks of a motor skill. Today, we have Jörn Diedrichsen.

Diedrichsen resumed the case for Lashley's lost motor engram. It had been 40 years since University of Southern California psychologist Richard Schmidt proposed a "schema" theory of motor command, arguing that some movements could be run automatically, without the assistance of feedback. His concern was storage. In order to produce the sounds required for an English vocabulary, Schmidt noted, researchers estimated that more than 100,000 phonemes are required. Are all those lip, tongue and throat movements held independently? The cognitive holding capacity would seem astounding. But so, too, would the demands on a central nervous system that has to constantly produce chains of reflex actions regardless of context. A simple task like screwing in a lightbulb—the result of perhaps a dozen distinct motor elements—would be far too taxing if some form of learning and consolidation into memory did not occur.

Even if there was consolidation into memory, how the skill manifests itself as a behavior is of equal relevance. Downhill skiers frequently transition into snowboarding. The skills are distinct but reliant upon several of the same characteristics. Yet the brain does not have to rememorize every body position in order to learn to achieve what is roughly the same task. If the brain did exert autonomous command over every muscle required for any movement, while monitoring all the incoming sensory input as well, such skill transfer would seem impossible (or at least much more time-consuming) to execute. Nikolai Bernstein recognized this. Grappling with the problem of our body's many degrees of freedom, he noted the topology, or what he called the "deeply rooted indifference," of motor control when drawing a circle, large or small, on a horizontal piece of paper or a vertical blackboard. "The almost equal ease and accuracy with which all these variations can be performed," Bernstein wrote, "provide evidence that all the variations are ultimately determined by one and the same supreme controlling engram." He called it his principle of the "lines of equal simplicity": "If a circle is drawn with an arm directly in front of the person, then directly out to one side, and then about some intermediate axis, both the muscle and the innervation schemes of the three movements will be very different. However, all three movements are subjectively very much similar in terms of their difficulty and objectively they show approximately the same indices of accuracy and variability." Bernstein theorized that the brain stored commands not as discrete muscular instructions, but as a more generalized spatial pattern. The motor engram was more like a gestalt formulation of stereotyped movement parameters: speed, trajectory, torque, distance. To us, the motor engram was more like the embodied representation of a goal.

With this, Bernstein laid the groundwork for the taxonomy of action. He did not think of movement as caused by specific motor commands. His concept of sensorimotor integration involved diverse regions of the nervous system—interconnected sensory and motor cortices, overlapping representations of the periphery, "the mutual activity of entire systems or organs which, anatomically and clinically, display varying degrees of independence," he wrote. This was not Sherrington's reciprocal innervation—it was multiple innervation, a far more active and cooperative reimagining of the nervous system and the body.

There is much to compare between the works of Lashley and those of Bernstein. And yet the Russian formulated these giant scientific opinions with limited access to Western works. And to the West, and to the West Virginian, Bernstein, who published many of his most influential experiments in the 1930s and '40s in the Soviet Union, was largely unheard of until the 1960s. Bernstein seemed driven simply to call attention to the flawed reflexology constructs of his own comrade, Ivan Pavlov. This view, the chaining of reflexes, which otherwise came to be known as the Peripheralist theory, considered a movement as a string of reactions triggered by a stimulus. Bernstein proffered a competing claim. He called it the "comb" hypothesis, and you can see why:

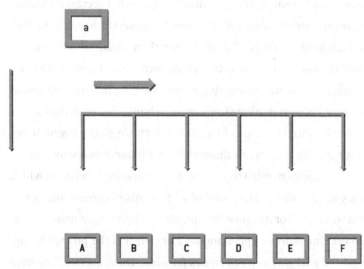

How Bernstein diagrammed one of his concepts of a motor scheme (such as the motions required to change a lightbulb). The engram or "motor problem," (a), prompts a successive chain of often smaller independent movements (A, B, C, D, E, F).

The top-down model helped launch a more hierarchic view of the organization of motor control, which Bernstein eventually described in five discrete levels. "Actions are not simply movements," he said. "Most of them are whole sequences of movements that together solve a motor problem." Bernstein's motor problem (a) is, by others' definitions, the motor image, the movement program, the scheme or the *Bewegungsformel*. It is the guiding engram, the target for the desired state of the environment. There is still a chain, just as he said lighting a cigarette involves a sequence of a dozen independent movements. But they are linked by the goal—a butt alight—rather than the trigger of some initial stimulus.

How they get linked is also important. One of the key aspects of Bernstein's framework is that the individual motor elements can still be integrated. This is the idea of chunking. For certain problems, like a memory test, for instance, chunking is critical in order to recall long sequences of numbers. We generally have a hard time remembering strings of digits longer than seven, which is why a 10-digit phone number is broken up into three digestible chunks. Anders Ericsson's earliest studies on expertise involved champion memorists who fashioned mnemonic chunks out of strings sometimes hundreds of digits long. But in fact, we chunk instinctively, not just with memory or imagery but with motor tasks, too.

Diedrichsen tested this notion by giving subjects a task in which they needed to tap a fingerboard with numbers corresponding to a certain digit. For instance, the thumb was 1, the index finger, 2, and so on. The number appearing on a screen was the prompt to tap, and the goal was to go as fast as possible. The sequence they were given was lengthy, but there was a pause in between prompts after every three or four digits. So the participant was being subliminally fed the chunks. The resultant patterns looked something like this:

1-2-4 | 4-2-1 | 1-2-3 | 2-3-1

After enough repetition, the 1-2-4 gets stored in their brains as a sequence that will *automatically* trigger the 4-2-1. After two days, Diedrichsen would remove the pauses in the prompts and have them continue to train the same sequences for three weeks. The participants became much faster. Yet, they continued to pause— ever so slightly but quantifiably nonetheless—at the portions where the chunks used to be assigned. They were being paid to go as fast as possible. Still, they paused to chunk.

Now these sequences were essentially hardwired into the brain. When enough sequences get accumulated, this forms a hierarchy, a comb or a pattern of motions, such that even the very initial sensory prompt will spit out the entire chain of sequences almost reflexively. You can think of it as the first note of the piano sonata—if that key is off, the whole piece is vulnerable to crumble. When Diedrichsen looked at this in the fMRI, in fact, using a novel analysis technique, he saw that parietal and premotor regions showed clear evidence of the encoding of these instructed chunks. In contrast, the primary motor cortex seemed mostly concerned with single finger movements. Interestingly, however, the first finger tap of a sequence was much more pronounced (as much as 40 percent more) than what followed, even in an 11-digit sequence. The first tap left a deeper imprint on the primary motor cortex. "It's like the first thing the motor cortex does is more important than the other ones," he said. In a satellite presentation at the 2016 Society for Neuroscience conference, Diedrichsen and a Japanese colleague, Atsushi Yokoi, deemed this to be the first neural evidence of the hierarchical organization of chunks. The giveaway, he said, was the clear evidence of the first finger movement.*

His discovery seems to point to the existence of what psychologist Jack A. Adams called a "memory trace" and "perceptual trace" in 1971. Something initiates the movement, prompts its early direction and sets it on its path. From there, experience (the perceptual trace) steers the movement along, comparing incoming feedback with what it expects based on memory. A 2015 study involving a technique called "optogenetics" seemed to confirm this concept a different way. By inhibiting the motor cortex—rather than electrically stimulating it, like Ferrier—a group from Howard Hughes Medical Institute found that mice froze their front limbs if the technique was employed midway through a skilled task like reaching for a food pellet. When they switched off the inhibitory signal, the mouse would suddenly make a grabbing motion, even if there was no food in front of it, as if it were completing a suspended thought. The fact that the mouse suffered none of these effects if the task at hand was simpler, involving routine instead of skilled movements, gave the authors reason to believe that the cortex either represents the seat of the engram, or "could trigger or permit its activation in a downstream network."

The engram may be just the goal of a trained behavior, and

---

* The concept of generalized motor programs is widely accepted, but not by everyone. A competing view, called dynamic pattern theory, arose in the 1990s and remains staunchly defended by some. Diedrichsen's fMRI results, however, will undoubtedly factor in the debate.

what gets the first finger moving is really the key. We have plenty of behavioral examples to illustrate how we kick-start ourselves into action. Many golfers require what they call a "swing trigger"—a slight knee kick-in (Gary Player) or a quick head turn (Jack Nicklaus)—to instigate their backswing. It may be to reproduce the hierarchy of their habitual movement. It may be why Daniel Wolpert clicks his tongue. It may be why some athletes are so fanatical about their routines. If, at the free-throw line, that first dribble is mistimed, the motor chunking after hours and hours of repetition could collapse. It is not purely superstition (although the pattern of the dribbles that follow might be). In the brain, in the primary motor cortex, the drive to trigger the first movements is extraordinarily strong. "It's like 'Get out of bed, you lazy bastard,'" Diedrichsen jokes.

Both Schmidt and Adams, long ago, each looked at what is called the "warmup decrement," which implies that even well-trained skills are susceptible to temporary waning after only a short period of rest. Basketball players seated on the bench for too long during games generally miss a few shots before they get back in rhythm. If it's days or weeks in between games, we have come to expect there to be considerable rust. To combat this, some athletes will practice for an hour or more *on the day of the game* before they are ready to play. For baseball players, batting practice usually begins at least three hours before the first pitch. But why do we need to warm up before a performance? A robot Ping-Pong player (look it up) needs only to be turned on in order to immediately start functioning. Is our infatuation with warming up just for calisthenics, so the muscles get loose? Actually, experiments have tried warming up the muscles in other ways, without taking the court or field, and found that the athlete usually ends up looking rusty and performing poorly.

It is not just for calisthenics. A warmup is said to be more legitimately required for a cognitive recalibration, like tuning a guitar before a set. Even if the instrument has only been rested for a few hours, that time away from the performance stage is filled with unrelated activities, activities that can perturb the delicate dynamics of the system. The brain is not a sieve; it is a sponge. Even unwittingly, it soaks up everything around it. Neurons move and connections strengthen and weaken, form and dissolve, constantly, reflecting the plasticity that enabled us to learn the skill in the first place. A warmup might be required to reorient the brain, jiggling loose the engrained motor programs that, for a few hours, had been placed on hold. It is preparing the chunks. Other researchers, including John Krakauer, have similar thoughts about certain idiosyncratic tendencies of certain athletes, such as Rafael Nadal's strange habit of tugging at his shorts in between points, or Novak Djokovic bouncing the ball 20 or 30 times before a serve. These "ticks" could represent the same warmup protocol—a primer to the brain that a complex string of action, the one you have actually been training for, is about to unfold.

Chunking is evidently beneficial to sequence action and complex movement (or memory tasks). But what determines the composition of those chunks, and how do you know if it is good or bad? The consolidation of motor actions into motor memories, as you already know, has generally been viewed as a positive thing. Unless, of course, you are simply reinforcing bad habits. A few years ago, the San Antonio Spurs tried to address this by hiring a shooting coach named Chip Engelland. It might sound a bit alarming for a successful NBA franchise to pay a consultant strictly to help its players reconstruct their shooting form. But that is exactly what Engelland did. His work, specifically with guards Tony Parker and Kawhi Leonard, helped break the bad habits that had been reinforced over years of practice.*

This is not simply a technical problem easily mended with conditioning or flexibility. It is wired into the brain. But why? Together with his graduate student Nicola Popp, Diedrichsen looked at his sequences and noticed some patterns. There were some transitions between chunks that were easier to produce than others. A transition from 2 to 3, for instance, involved taps of the index and middle fingers. That's easy. That was something the participants could do relatively quickly. But when 5 was involved, especially if the pattern went 5-5, repeated pinkie presses are hard. It is difficult to reproduce the requisite amount of force, and therefore participants usually have to pause to do it. So Diedrichsen rewrote the pattern to break down into chunks that he thought would naturally require pauses anyway. For one group, the pattern went something like this:

1-2 | 3-5-1 | 3-3-2

And for another group he gave a pattern of chunking that was not so ideal:

1-2-3 | 5-1-3 | 3-2-1 | 3-4

The bad pattern actually cost participants 200 to 300 milliseconds compared to the good pattern after two days. Even after three

---

* Leonard, a 25 percent three-point shooter in college, has become a 38.8 percent three-point shooter in the NBA and a consistent contender for the league's Most Valuable Player award.

weeks, the cost remained 80 milliseconds. That might sound like nothing, but on a sequence that takes only 1.8 seconds to execute, the cost is about 5 percent. Think about that for a downhill skier or a young musician trying to strike chords on a guitar. If the brain cannot compensate for the inefficiency of a learned sequence in a simple skill like finger taps, what might happen if bad habits are formed in something far more complex? These "chunks" were manufactured, and then Diedrichsen discarded them. He removed the breakdown of the sequences. The participants could have rearranged their taps in any sequence they wanted. Yet they continued to tap in a pattern that delayed their performance. Their brain had chunked the sequences into a hierarchy, and that hierarchy was difficult (although not impossible) to break.

Herein, then, may be that lesson (or a warning) again for coaches about the right proceduralization for the right tasks. "Some ways of chunking are biomechanically better, can lead to better performance," Diedrichsen says. "Once the system picks up one way of conceptualizing it, it's very hard to get out of it. They don't spontaneously get out of it." Bad habits are very hard to break. The trickiest thing about them is that the participants' scores did improve, so it is not as though the habits completely hampered their performance. It hampered it only a little bit. About 5 percent. How many amateur athletes would even notice that? The knowledge might not take somebody with a 10 handicap and magically transform him into a professional. But without proper coaching, it might explain certain plateaus that many of us reach and struggle to overcome. "You put the system already on that track," Diedrichsen says, "and they're going to keep going down it." As Warren Buffett likes to say, "The chains of habit are too light to be felt until they are too heavy to be broken."

Chunking has its pitfalls and its positives. For a musician, the flexibility of the arrangement of chunks can be what distinguishes a great improviser from someone who requires more structure. A jazz artist like Charlie Parker could take elements of stored saxophone riffs and splice them together interchangeably. He could listen to what he was playing and decide in that moment the next chunk he wanted to include. This would not work for a guitarist attempting to memorize the precise licks of "Purple Haze." But it helps for those instances when a sequence you have practiced looks like something you haven't—such as riding a bicycle with the handlebars reversed. If you have wondered—as I often have—how in the world Michael Jordan adjusted his soaring body to transform a right-handed slam dunk into a soft left-handed layup in an NBA Finals game against the Los Angeles Lakers in 1991, the answer might not only lie in his genetic gift of hang time. Like Parker, he took one riff and spliced it together with another. For most of us, an on-the-fly attempt like this would have ended in embarrassing fashion. But not for MJ. His ability was predicated on skill, not habit. The representations in his brain were ready to make that transition look as easy as a flip of a switch.

There is evidence, too, that even when we are not intentionally practicing at something, we can still be passively absorbing it. *The brain is not a sieve; it is a sponge.* In learning to play the piano, we train in chunks that require our hands to work independently of each other, suppressing the tendency to make mirroring movements with both hands. Over time, this enables the hands to work the keyboard in distinct ways. But there is growing evidence that both sides of the brain are working hard to learn the sequences trained for the *opposing* hand, even if that action will rarely manifest itself in behavior.

In the spring of 2012, while at University College London, Diedrichsen and a colleague, Fred Dick, along with PhD student Sebastian Telgen, sought to understand a little more about learned hand sequences by inviting amateur musicians into the lab. These were not just any group of amateurs—they were highly trained violin prodigies from the Royal Academy of Music. Dick, a violinist himself who trained at a conservatory in Germany before turning to science, helped design the experiment, which was intended to push the experts far out of their comfort zone. Accomplishing this inside the bore of an fMRI scanner is not easy, but Diedrichsen and Dick did not have to push too far. They simply asked the violinists to manipulate the strings of the violin with their right hand instead of their left. Yet it was as if they had asked them to sign their names holding the pen between their toes.*

Violin is a funny thing. Both hands have to be "coordinated but uncorrelated," Dick told me. If you watch trained violinists play the piano, "they generally suck. Especially with the right hand. Even though most people are right-handed." The right hand has just spent years learning to operate the bow as it glides across the strings. The chunks for that movement have been well established. Same goes for the left hand, which relies on the fingers to apply more ballistic

---

* This study has yet to be published, because getting access to the violinists has been difficult. And then Diedrichsen left UCL. But they hope to eventually publish.

movements to stop the strings. And so, knowing all this, Diedrichsen and Dick asked the experts to switch-hit.

Telgen had fashioned fake fingerboards using real violin strings that could work in the scanner, and the task was to play the first page of the first book of Schradieck's *The School of Violin Technics*, a warmup exercise the experts could all but do in their sleep. When told to switch hands, what Diedrichsen recalls is that none of the participants were very pleased. "They said, 'No, I cannot possibly do that.'" And then they got into the scanner and did it. They did it well. They did it *fast*. Some were faster than the controls—those with little to no violin experience—using the correct hand. All those years of practicing with the left hand had inadvertently transferred to the right.

Dick was stunned. "I can imagine as a baseball player, I can at least hit a ball batting left or right," he told me. "No fucking way I could play the violin at all [with my right hand]. I think I might drop it." That is what everybody thought. Get in the scanner, though, and those sequences somehow materialized. "People were really, really fast and showed massive transfer of the learned patterns to the right hand," he said. "That surprised everyone."

Diedrichsen's understanding of the functional organization of the brain did not allow him to be quite as surprised. Recent research has begun to reform that long-held wisdom about the contralateralization of our hemispheres, which, Shlomi Haar of Ben-Gurion University of the Negev told me, was based largely on "preliminary, simple experiments" that were "actually not capturing the full story." Like Diedrichsen, he has found relationships between the brain responses of one arm and the brain responses of the other, suggesting a lot more generalization of skill learning across the two hemispheres. The nondominant arm might not appear as "skillful," but that hardly makes it useless. When we hammer a nail into the wall, the nondominant hand is there to stabilize the nail being struck. Or holding a plate of food, we can balance with one hand while we shovel on spoonfuls of potatoes with the other. A focal dystonia in one hand has been shown to appear in the other while performing the same tasks. "The nondominant hand, though it's weaker and less skilled, is still perfectly coordinated with the dominant arm," Haar said. "You have a very strong relation between the two."

It is Diedrichsen's belief that transfer does not occur during encoding or retrieval. It does not *occur* at all. It is a fabricated step in the generalization process that semantics has added. The brain does not work the way Sherrington conceived it, or the way Penfield conceived it, as a strict and unitary flow of coordination. The motor system is bilateral, even if the output does not conform to it. Our hands are smarter than we think.

Penfield was right about something. If you look again at the homunculus depiction, the hand is huge. It stretches almost a full quarter of the map. It is like an early sketch of the Louisiana territory. The same is true of Penfield's subsequent drawing of the sensory homunculus, which plotted the most sensitive segments of our bodies. The fingers alone take up about twice the brain area of the elbow, forearm and wrist combined. There is more cortex devoted to the fingers than to the entire trunk of our body. The sensitivity of our fingertips is not just translated into more cortical surface area. They, along with the tongue and lips, contain the highest density of sense organs—called "mechanoreceptors"—on the body. About

2,500 reside on each fingertip. When some object, like a baseball, touches the hand, the skin conforms around its contours, creating a mirror image of the surface of the object. This contouring stimulates the four types of mechanoreceptors found in the skin, each of which provides different inputs to the brain about the object: (1) its shape and (2) surface texture, (3) how much force is being applied to it, and (4) how much it might vibrate in our hand. This allows us to feel the dirt as we plunge into it with a shovel. Blindfolded, most adults need only to feel 1 millimeter of an object on their fingertips in order to adequately discern what the object is. When we grasp an object, we need even less—just 0.5 millimeter—to discriminate features of its surface. Tactile acuity is slightly better in women than in men. It varies between fingers but not between hands. The distal pad of the index finger is the most sensitive. The palm is six to eight times less discerning than the index fingers. Two objects held to the skin of my wrist would need to be placed at least 40 millimeters away from each other in order for me to discern they are different. In my fingers, I can tell the difference within 2 millimeters, which is why we don't attempt to read Braille with our elbows. Anyway, it might look funny.

The quantity and quality of these mechanoreceptors are what not only enable our dexterity—knowing where our fingers are in space—but allow us to manipulate the objects in our grasp, whether they are chess pieces, a pencil or the seams of a baseball. Andrew Pruszynski, another professor at the University of Western Ontario, has refrained from referring to his work as "motor control" anymore. "I say 'sensorimotor control,'" he told me. "You can't separate motor from sensory in any simple way. They're intrinsically connected. In fact, when we record from neurons in the primary motor cortex, there's a lot of sensory information being driven through those neurons as well. It really looks like a close coupling between sensory and motor is key for successful motor control."

I could understand where visual acuity had a role in the ability to hit a baseball, but the tactile acuity of holding the bat or feeling the mound beneath the cleats? I hardly saw the connection. But he convinced me to look again. So many of our most impressive physical capabilities—outside certain performance art forms like karate or dance or running—involve both the delicate and forceful handling of objects, sometimes interchangeably, in a manner uniquely human. When we hold a pencil or a rake or a baseball bat, studies have shown that the neurons in the parietal lobe actually adjust our visual receptive fields to extend to the end of the tool. It is why we can use chopsticks to pick up fine grains of rice and why drivers believe they can sense the road as they steer a car. This amazing capacity for the brain to morph the perception of its own body to incorporate a lifeless object is also reliant upon something: familiarity with the object. The brain needs time to adapt, whether you are first learning to walk with a cane or dribble a hockey puck at the end of a stick. It needs practice.

And then, maybe, there is another factor. Sensory input is a critical variable in any Bayesian equation for human motor behavior. In his computational descriptions of motor control, Daniel Wolpert placed a high value on the expert's attuned relationship with his senses. "I'm sure there are ways to hone both your sensory abilities and your motor abilities," Wolpert told me. "It's not just your motor output—you need to be able to sense really well. You have to be fast in your sensing."

I assumed Wolpert was referring to those perceptual abilities we would expect: mainly vision, maybe some hearing. But Pruszynski was actually not talking about those senses at all. He was talking about touch. Could there be differences in how we feel?

In his office, Pruszynski pulled up a video that would demonstrate to me the importance of haptic perception. I thought I had a good idea of how important it is. My foot falls asleep more frequently than I would like. In this video, though, a young lady was asked to light a match, which she was able to do with ease. Then they anesthetized her fingertips with lidocaine. Only three fingertips on her right hand. What's more, the anesthesia was fine enough to block the nerves from just some of the mechanoreceptors (and remember there are thousands) in the skin of the tips of those fingers. She has all the vision in the world. She has nothing blocking her ears, she is well practiced and she knows the task. And yet . . . fumble. She drops the match several times. At one point, she does not even know the correct way to orient her *hand*. It takes nearly a full, excruciating minute to light the match. And this, Pruszynski points out, had nothing to do with motor control. "Not even the muscle receptors are problematic," he says. "It's only the receptors in the skin."

Those receptors, whose neurons connect directly to the spinal cord along a single, long axon, are incredibly sensitive and remarkably conductive. With microneurography, a technique that allows for the recording of nerve impulses on the arms or legs of awake human subjects, researchers can focus specifically on the neurons that innervate the fingertip. In other parts of the body, it takes a lot for the brain to notice the skin is sending signals. When 1,000 neurons are stimulated, sending signals to the brain, you might feel that. But, amazingly, it's been shown that you can perceive when just *one* fingertip neuron is stimulated. "That's the quanta of information you have in the nervous system," he said. "There's no smaller unit of information that you can transmit." Yet that piece of information is potent enough to be perceived in the brain.

While electricity can travel through a copper wire at 96 percent the speed of light, or approximately 645 million miles per hour, the transmission of electrical currents from the skin's mechanoreceptors takes place at an average of about 150 miles per hour. In his book *Touch*, the neuroscientist David Linden relates this interminable delay to how it would play out in the body of a giant sprawled out across the Atlantic Ocean. If seaweed along the coast of South Africa brushed her big toe on Monday at noon, she would not feel it in her brain in Baltimore until midafternoon on Wednesday. The rate of communication between the skin and the somatosensory cortex is akin to snail mail in a wireless world.

What to do with that significant lag? The body adapts. In fact, in an astounding revelation not long ago, Pruszynski found some cases in which our sense of touch might not even need the brain at all.

As afferent nerves lead out from the spinal cord to the distant periphery of our hands, neurons are densely packed into the skin. These neurons have axons that feather out into mechanoreceptors along the surface, producing expansive and highly sensitive receptive fields, like a web catching perturbations from beyond the exterior. And yet these zones are not uniformly distributed. If a pinprick lands on the entire web, the neuron will fire a great number of signals up to the brain. But if something touches only a few of the web's mechanoreceptors, the signal frequency might actually be different. In a paper in *Nature Neuroscience* in October 2014, Pruszynski and Roland S. Johansson, from Umea University in Sweden, wrote that they believed this to be a kind of code.

What sort of stimulus would get only a few but not all mechanoreceptors on our fingertips to fire? The edge of a surface, they surmised. Indeed, the skin itself might be able to determine the orientation of an edge without the brain's involvement at all. It is as though your fingertips are making calculations on their own. "It's not that we're saying, 'The neurons on the skin do everything,'" Pruszynski told me. "But they start the process. They preprocess it." Edge orientation, he continued, is critical in motor control. Our daily lives depend on knowing precisely where smooth surfaces begin and end, enabling us to button a shirt, grasp a cup or grip a railing.

When I called Linden at Johns Hopkins to hear what he thought about this, he said the finding was not wholly surprising. There are also sensory nerve axons, he said, called "C-tactile fibers" or "caress sensors," that are actually designed to carry information about interpersonal touch. There is such a thing as "optimal" speed for a caress: too slow and it might feel like a bug crawling along the skin; too fast and it doesn't carry the appropriate amorousness. "We now know," Linden said, "that if you stick an electrode in the arm and record electrical activity from one of those C-tactile nerve cells going from the skin to the brain, the highest firing rates happen at precisely the touch velocity that 'feels' best to us." The encoding of that optimal rate is not done in the brain. It is done at the periphery. "The properties of the sensory end organs are already conditioning what is important before the brain even gets involved." Our skin is telling us something.

It may be plausible, therefore, to conceptualize a schoolyard where some children have a better grip on the pebbled rubber of a basketball than others, based purely on the firing patterns of their mechanoreceptors or the sensitivity of their caress sensors. The yo-yoing dribbling ability of NBA guard Kyrie Irving may not be so incomprehensible to fathom. How would we know if some people "touch" differently? "Feel" is a term bandied about in locker rooms or on sports broadcasts where announcers use it in empty expressions like "feel for the game" or "feeling it." A good putter in golf should have good control over his hands and wrists, but the tactile sensations of the club as perceived by the brain are not exactly simple to quantify. It is also reasonable to wonder what difference it could make to a motor performance whether or not your sense of touch is as good as the next person's. Would that really be so limiting? Well, consider that the slightest decrease in air pressure inside the footballs used by the New England Patriots in a playoff game against Indianapolis in 2015 set off a firestorm that would embroil quarterback Tom Brady in a lawsuit against NFL commissioner Roger Goodell, lasting for over a year. Brady clearly likes the grip on his footballs to be a certain way when he plays—enough that there were suspicions he tampered with them to conform to his preferred specifications. Green Bay Packers quarterback Aaron Rodgers, on the other hand, has said he prefers the football to be *overinflated* to his liking. Their skin might be telling them something. How much is unknown, but it would be a mistake to consider their grip to be a trivial variable in the overall execution of their motor system.

Aristotle did not overlook the skin. "The human being is left behind by many of the animals," he said, "but with respect to touch he is precise in a way that greatly surpasses the rest, and this is why he is the most intelligent of the animals." He was wrong—our tactile perception is no more sensitive than many other animals'—but his interest in the sense of touch was extraordinary when compared with almost every researcher interested in studying the brain up until this century. Touch, in comparison with the other senses, has been shamefully neglected. For about every 100 studies on vision, there has been one on touch, Linden says. The proper role of haptic input is still years away from being mastered in robotics, which is why you have yet to see a robot capable of fishing around for its keys in the pockets of its jeans. Researchers at Ben-Gurion University in Israel in 2010 offered $1,000 to whichever team could design a robot that was simply capable of shaking a human hand to an *approximately* human degree. They gave the teams a full year to do it. The winning team incorporated a mechanical telerobotic interface—sort of like a handle that moved up and down—that would hardly be confused for Rosie of *The Jetsons*.

We all know how it feels to momentarily lose our sense of touch in those fuzzy moments when a foot or arm falls asleep. But the impact of sustained touch deprivation can yield dark consequences. In the understaffed Romanian orphanages of the 1970s and '80s, under Ceaușescu, babies were born and left to grow without the tactile engagement infants require—the warm hugs and soft caresses that, it turns out, can be of lasting influence. Through adolescence, many of the children sustained not just emotional or psychiatric issues, but physical defects as well: stunted growth, gastrointestinal disorders and a reduction of gray matter in the brain.* There is also

---

* John B. Watson, one of those forefathers of the Behaviorist camp, used to advise parents not to touch or hug their children, which he perceived as promoting weakness. We recognize now that Watson could not have been more misguided.

evidence of the power of touch impacting our daily performances in adulthood. In 2010, researchers at UC Berkeley undertook a study of teams in the NBA during the 2008–09 season to measure whether the number of high fives, chest bumps and butt slaps among teammates in a game early in the season could predict the future success of the team and how cohesively they played. As it turned out, the answer was yes. They found that players who were more embracive of teammates had higher "win scores," which measured positive impacts on a team's success, and the league's best performing and most selfless teams were also its touchiest. Not surprisingly, some franchises took notice: In 2016, the Phoenix Suns introduced a "high five stat" to measure how many hand slaps players were giving one another throughout the game.

It is instructive to remember that there is more cortex devoted to the fingers than to the entire trunk of our body. Several studies, including one involving professional violinists and cellists, found that representations in the primary somatosensory cortex of the fingers of the players' left hands (used for the tactile stimulation and manipulation of the strings) were almost twice as robust as those for the fingers on the right hand. The brain responds to input, and the more dexterous digits were clearly producing dynamic changes to the cortex of the brain. It does not just have to have a motor component: As women breastfeed, the mental representation of the nipple increases, and later returns to normal size. The touch map can grow and shrink no matter how passive the activity. We also know that the skin loses its perceptual ability over time, as mechanoreceptors gradually thin out and the rate of transmission of electrical impulses naturally retards. From age 20 to age 80 our fine spatial acuity will have decreased by half. This is partly why we seem to lose our balance more as we age: The touch receptors in our feet diminish and we have more trouble feeling the ground. The balletic movements of pitchers on a Major League Baseball mound require an inordinate amount of balance that often goes underappreciated. That is crucially dependent on the touch receptors in their feet. It is hard to imagine a gymnast in sneakers deftly cavorting along a balance beam.

The jury is still out about the true relationship between the sensory and motor systems, but Pruszynski is ready to deliver his closing argument. "You don't really have motor without sensory," he says. In some cases, you don't have sensory without motor. Tactile receptors are fairly marginal if you just sit there pressing indentations against your skin. But those receptors really kick into gear if the skin is in some form of motion. We might consider how Pruszynski likes to think of the influence of haptic feedback on an expert pianist. You could imagine there being two possibilities: As you get really skilled on the piano, you're just so good motorically that the sensory inputs become irrelevant. You're like a robot. Another possibility is that what makes you great at piano is that you're *more* attuned to the sensory inputs that guide your next action. A piano key, after all, is an inch-wide rectangle. You have to press that key and know where exactly to move the finger to the next key. To do that, you have to take into account where you landed on the first key and with the right amount of force. That takes an enhanced perceptual capacity. Perhaps there is even an in-between: People who get to be good become robots—they tune down the sensory information that they needed when they began learning. But then, to become really, really good, the sensory strategy reemerges in a different way. Your performance becomes more sophisticated. You pay close attention to things like the pounds per square inch of the football, because it can impact your performance.

"It's in the details where somatosensory really shines," Pruszynski says, "and that's where it has been overlooked."

7.

EMBODIED EXPERTISE

WATCH AND LEARN

Sometime not long ago, Kerry Spackman splurged on a hand-made Lotus Seven, a replica of the 1960s-era British roadster that resembles a steel cigar with bulging eyeballs. The wheels jut out and the engine rumbles, as if a heavy trash can were being rolled along a gravel driveway. He insists the car is street legal. But Spackman does most of his driving on a track near his home in Auckland, where he can properly unleash his racing spirit. Spackman says that the car regularly outperforms other two-seaters of its day, such as those made by Ferrari or Lamborghini. But he cannot be sure quite how fast he takes it, because he is afraid to take his eyes off the road.

At the very least, he knows that his car can reach 60 miles per hour from a standstill in 2.9 seconds. The g-force, or the amount of times the force of gravity is felt at a particular moment, on that type of acceleration registers 0.97. When Formula 1 drivers brake suddenly around a bend, the g-force can reach 8, causing blood to flow rapidly toward the head and straining the limit for human tolerance before loss of consciousness. Racers call this "cornering." Fortunately, modern F1 cars don't have much need for brakes anymore. They can hug bends at speeds upward of 200 miles per hour, with cornering maneuverability likened to that of fighter jets more than stock cars or other automobiles. Spackman is a racer at heart, but even he understands his limitations. "I don't have the innate hardware," he told me, "in my brain."

Instead, he has studied those who do. In the early 1990s, Spackman, whose focus on applied mathematics at the University of Auckland decades earlier helped him develop and sell a patented radio-tracking device for race cars to Ford Motor Company, was introduced to Sir Jackie Stewart, then assisting in the testing of Ford's race cars. In addition to winning three World Drivers' Championships, and twice finishing runner-up, in nine seasons between 1965 and 1973, Stewart was a star marksman in his youth, winning the British, Irish, Welsh and Scottish skeet-shooting championships. He had an extraordinary memory capacity for the smallest details, from the angles of corners to the tiniest idiosyncrasies of the car. He was also dyslexic. Whether any of this contributed to Stewart's preternatural ability behind the wheel fascinated Spackman, and likewise intrigued Ford, which was interested in finding the next Jackie Stewart. "Ford said, 'How do we get everyone as good as Jackie?'" Spackman recalled. "I said, 'I have no idea.'"

Spackman has become sort of a race whisperer for many of the sport's top drivers, but he started by analyzing Stewart. He began with a simple measure of reaction time. Stewart, who had retired two decades earlier, in 1973, was not exactly in top form. But it was hard to say whether he was ever quickest among his peers. As Spackman expected, Stewart's reaction time was nothing noteworthy. So in a sport of mind-bending speed, in which traditional "athleticism" is constrained to minute adjustments of essentially only the wrists, hands and feet, what separated a star like Stewart from the rest of the pack? Formula 1 is often described as a test of both man and machine, and certainly some cars perform better than others, but the consistency and dominance of Stewart throughout a golden age of the sport must have meant he had some gift that others lacked.

One day, Spackman decided to sit Stewart in a rotary chair designed to test the vestibular system and run him through patterns. He was instructed to yell whenever he felt the chair move. Some of the tiniest movements were practically imperceptible. Yet Stewart recognized every one. What's more, after the exercise was over, he was asked to recall the positions into which the chair readjusted. Stewart could recall every fine detail. They brought him onto a racetrack for a double-lane-change drill, in which the driver must swerve to avoid an object. Electronically, Spackman could make adjustments to the vehicle from afar. Stewart sensed it all. When it was over, his recap stretched on and on. "As I first started to turn the steering wheel five degrees, I'd notice the left front bushing had a bit more compliance," Stewart would say. The actual lane-change maneuver elapses over 1.5 seconds. Stewart could analyze it for 10 minutes. "We looked at the telemetry data and everything he said was there, captured in incredible detail," Spackman said.

Stewart seemed to possess an extraordinary capacity for not only perceiving sensory inputs, but encoding them, storing them and then recalling them (as memories) at a moment's notice. Spackman was at first astounded. But it began to make sense. Stewart was considered a clinician on the roads. Most Formula 1 drivers say that they think two corners ahead, but Jackie would analyze each one at a time. In his mind, turns did not consist of only three main portions—the entry, apex and exit. He would break them down into eight, with each point constituting different afferents he would see or feel. He could recount them with stunning detail. It was not so much about replicating the maneuvers, but about understanding the differences required to maximize his speed from one turn to the next. Spackman now had the earliest clue as to how this was possible. Stewart could go around a corner at 100 miles per hour and still observe 20 more things than you or me. He was also able to remember all the nuances for the next lap as well, to use them for comparison. "Eventually, he can hone in on an optimal solution," Spackman said. "It's almost like he sees the corner through a microscope." Spackman added, "He was in a class of his own."

Spackman began working with Formula 1 teams, including McLaren and Ferrari, in 2001. Because the sport prefers to keep its advantages behind closed laboratory doors, he never published any of his research, despite access underneath the balaclavas of some of the foremost driving experts in the world. But he did gain extraordinary insights. Working with the driving simulator, in particular, Spackman used to enjoy finagling with the peripheral flow field—the space that flies by as the driver gazes ahead—to gather understanding about the hardware that defines an expert racer's ability. The racers somehow managed to "make use of that peripheral information," more than a nonexpert, Spackman said, "and they're more precise at it. They're not even aware they're doing it, but it's happening anyway."

Spackman's work predated Jason and Jordan's by about a decade. But in many ways, what they discovered about the "innate hardware" of an expert in two vastly disparate fields was easily relatable. The baseball hitter is not reacting to the pitch by some sort of reflex action, just as Jackie Stewart was not simply using corners as slingshots. They used their experience to guide predictions that could better inform their actions. They moved to see and they saw to move. Jason and Jordan thought they traced that ability, in the hitters, back to portions of the cortex, the fusiform gyrus and SMA, which seemed to help them recognize and hold off on pitches that others would swing at.

Our sense of wonder when watching elite performers often arises because of not just their physical endowments but also their inimitable talent for making the right decisions at the right time. When a Formula 1 driver, for instance, applies the correct amount of acceleration at the correct moment of a turn. When Sidney Crosby threads a pass to a teammate. When a soccer goalie leaps to intercept an incoming shot. The movement is triggered by something the athlete sees, or senses, that others, even those on the same field, do not. This capacity helps us think of elite athletes as fundamentally different from others. But all of us contain certain links between perception and action, seeing and doing, which have evolved along with us. Without them, we never would have managed to avert predators, seize prey or jump out of the way of an oncoming bicycle as we crossed the road. What makes their connection seem so much better?

In 2008, researchers at Sapienza University of Rome studied a group of professional men's basketball players as they viewed film of a teammate shooting free throws. The film was fixed to cut off at the moment the ball left the player's fingertips, leaving the participants to guess whether he had made the shot or missed it. As it turned out, the players predicted the success of the shots correctly 66.7 percent of the time. It was an astounding rate—more than 26 percent better than coaches and seasoned basketball journalists. The researchers also applied transcranial magnetic stimulation (a method of measuring the activity of brain currents) of the left primary motor cortex while the observers watched the clip. When the shots missed, corticospinal excitability was also higher in the pros, specifically for the right hand muscles. It seemed there was something special about *active* expertise that the athletes not only saw but also, in a way, felt. The lead author, Salvatore Aglioti, surmised that expert players subconsciously ran a "motor simulation" of the action, which matched the observed actions with their own experiences to better predict what might occur. Even though the study involved other experts, it was the active players who seemed better equipped to make the most accurate predictions. Anticipation is a critical element of sports, where dynamic scenarios unfold quickly and actions must be made with sparse information. It is not enough to just bear witness to the events from the sidelines; you must be much more intimately, physically and actively attuned.

More studies tried other ways to confirm the results. In 2011, Aglioti and others found that motor experience gave a similar advantage to expert volleyball players, who performed better than non-players who had consistently attended matches for at least a decade, at predicting the outcome of a serve just by watching the body kinematics, even though the clip stopped just as the ball touched the fingertips. Another study used video of a soccer player making a penalty kick. A few years later, some researchers at the

Institute of Sports Science at Giessen University gathered 15 tennis experts and 16 volleyball experts from the highest levels of German sports and put each inside an fMRI scanner. They, too, were shown video clips of tennis serves and volleyball serves, with the clip interrupted just as the server was about to make contact with the ball. As they anticipated which direction the serve was headed, select brain regions lit up. The commotion was concentrated in a few select regions, such as the superior parietal cortex, broad swaths of the cerebellum and, indeed, the SMA. This is the primary acreage of a compelling and widely misunderstood causeway of neural activity. It is called the "action observation network" (AON), which sounds like a YouTube channel or another subsidiary of ESPN. But I like to think of the AON as a constellation, twinkling meekly against the backdrop of a dark and frenetic galaxy. It is the pattern that might signify the seat of perceptual expertise.

In 1890, the philosopher William James said the human instinct for imitation is unparalleled in the animal kingdom. "His whole educability and in fact the whole history of civilization depend on this trait." Humans tend to speak, walk and behave like others, usually without any conscious intention of doing so. There seems to be, James thought, a blind impulse to act "as soon as a certain perception occurs."

As far as James could tell, the brain-to-arm link was through the eye. He called it "ideomotor action," a name coined earlier by the British physiologist William Benjamin Carpenter, in which the mere thought or slightest perception of motion could trigger a corresponding movement in an observer. James saw no discrepancy between consciousness and action. "Every pulse of feeling which we have is the correlate of some neural activity that is already on its way to instigate a movement." If he noticed dust on his sleeve while talking with someone, he would reflexively move to brush the dust off without interrupting his speech. Sometimes ideomotor action did not need to prompt any motor output at all. Quoting Hermann Lotze, the German philosopher, James said, "The spectator accompanies the throwing of a billiard-ball, or the thrust of the swordsman, with slight movements of his arm; the untaught narrator tells his story with many gesticulations; the reader, while absorbed in the perusal of a battle-scene, feels a slight tension run through his muscular system."

Other psychologists were less convinced. That an idea can produce a similar action is "kith and kin with our forebears' belief that dressing to look like a bear will give you his strength," Edward Thorndike, a proper Behaviorist, wrote in 1913. Like the others, he dismissed the relationship between perception and action as subjective and, thus, unknowable. Worse, he impugned the principle of ideomotor action as an answer for psychology's "cravings for magical teleological power," along the same lines as superstitions, rain dances and voodoo dolls. From that blow, it was hard for the theory to recover.

But in 1952, frustrated by the field's general failure to view mental activities in any relation to motor behavior, the physiologist Roger Sperry advocated for a new approach. He argued that the literature needed to more thoughtfully consider motor patterns. "In a machine," Sperry wrote, "the output is usually more revealing of the internal organization than is the input." An examination of the finished product—the movement—might likewise be more enlightening than analysis of the ephemera that went into it. Sperry began to view the brain for what it really is: "a mechanism for governing motor activity" whose primary function was "essentially the transforming of sensory patterns into patterns of motor coordination." He sounded a lot like Daniel Wolpert. Sperry had looked in vain throughout the brain's architecture. There was nothing he could find in the cerebrum that was designed for anything but excitation of the final, motor pathways. Thinking, therefore, would seem only to be necessary for the refinement of movement, as well as direction toward future goals, adaptation and survival value. Sperry began to consider perception and action as interdependent and cyclical: One served to enhance the other. A new description for the meaning of perception emerged. When someone is looking at a drawing of a triangle, Sperry argued, his perception of it, his understanding of it, instinctively and automatically prepares him to respond to it. He is ready to point to it, trace it with his finger or move his lips to describe it. "This preparation-to-respond is absent in an organism that has failed to perceive," Sperry wrote. "Perception is basically an implicit preparation to respond. Its function is to prepare the organism for adaptive action."

Sperry's rethinking of the meaning and significance of perception helped revive the Jamesian way of thinking about the motor system's symbiosis with the mind. Other Cognitivists soon began the first investigations into the neurophysiological mechanisms underlying action and perception. Vernon Mountcastle, at Johns Hopkins in 1975, used techniques that enabled him to track the movement of individual neurons in the macaque. He focused on the posterior parietal cortex of the brain as the animal performed simple behavior actions (like pressing the key on a telegraph machine or pulling a lever) in response to sensory stimuli (typically a light shined in its eye or a subtle vibration). "Our results lead to a hypothesis of the function of the posterior parietal cortex," Mountcastle wrote. "These regions" receive signals that inform "the position and movement of the body in space, and contain a command apparatus for operation of the limbs, hands and eyes." It was the first time a particular brain area would be considered a throughway between perception and action.

Then in 1992, a team of researchers at the University of Parma in Italy decided to look at neural activity of a macaque, by training the monkey to retrieve objects from a testing box. Through single-unit recordings, they inadvertently noticed something extraordinary: Certain neurons, in the ventral premotor cortex, fired in similar ways when the monkey performed an action to when it watched one of the experimenters perform the same action. They recorded the firing patterns. *Pop pop pop.* "We incidentally observed that some experimenter's actions, such as picking up the food or placing it inside the testing box, activated a relatively large proportion of F5 neurons in the absence of any overt movement of the monkey," the authors reported. By 1996, after further tests, they had a name for the phenomena: "mirror" neurons. Monkey see, monkey do. James' assertion a century earlier about the animal instinct for imitation seemed to be empirically confirmed. The discovery was hailed as one of the most important in modern neuroscience.

The presumption followed that humans should possess a similar system. There was a seductive appeal to the concept of a biological mirroring system, one that made sense intuitively. We are, for the most part, visual learners. We are excellent at adapting someone else's movements for our own utility. We prefer to show, not tell. A description of the proper form for a golf swing is not usually as effective as attempting to mimic the movements of an expert. All of us, to some degree, are natural actors and imitators. But the presence of neurons with actual mirroring traits was not as easy to find in humans as it had been in the macaques. Neuroimaging investigations showed activity in the motor and premotor brain areas for movement preparation corresponding with observation of someone else performing the task. By 2006, there were studies involving highly familiar actions, or actions that are easily imagined, that supported the belief that the neural processes for observation and execution might indeed be entwined. But there was no conclusive proof.

That did not stop some science reporters from breathlessly chronicling the pursuit, or researchers across a wide range of disciplines from declaring ambitious conclusions of their own. By 2014, there were more than 800 papers published that involved mirror neurons. Neuroscientist Vilayanur Ramachandran compared the understanding of mirror neurons for psychology to what DNA meant for biology. The philosopher A. C. Grayling suggested that the evolution of mirror neurons was tied to humans' capacity for empathy. A "broken mirror" theory was established for children with autism (later debunked). The existence of cells that respond to other people's movements became aligned with everything from action understanding to mind reading to the popular success of romantic films to the shaping of human civilization to why yawning is contagious. Naturally, there was also pushback—mirror neurons were dismissed as the "most hyped concept in neuroscience"; "media-driven"; the "left brain/right brain of the 21st century"; and "neurobollocks." UC Irvine neuroscientist Gregory Hickok wrote a book called *The Myth of Mirror Neurons*, and Giacomo Rizzolatti, as a prideful forefather of the research, wrote a response paper defending his discoveries. A 2009 study by researchers at the University of Trento reported that there was no evidence of mirror neurons in humans. Then, in April 2010, a paper in *Current Biology* appeared with what it claimed to be the smoking gun: single-cell mirroring responses they had observed in 21 epilepsy patients as they executed and observed emotional expression and hand-grasping motions. Intracranial-depth electrodes implanted in the patients' brains allowed for access beyond previous noninvasive methods. The reaction, this time, was mixed. Many saluted the discovery as long-awaited confirmation after years of speculation. Yet others raised more doubts. They cited the fact that the cells in question appeared, among other places, in the hippocampus, an area not thought to be involved in the mirror system. The inferior parietal lobule and ventral premotor cortex were ignored because electrodes couldn't be placed there. Some neurons seemed to exhibit "anti-mirror," inhibitory properties, raising more questions. And reciprocal studies have yet to show that monkeys also have mirror cells in their hippocampus region, a reasonable parallel.

Around the time the wave of mirror-neuron mania was starting to crest, in 2005, a third-year PhD student at Dartmouth named Emily Cross clipped out an article on the subject from the May 12 issue of *The Economist*. She was also interested in investigating the connections between seeing and doing, and considered the mirror-neuron work as a fascinating potential clue. But something about it also left her feeling unsatisfied. She had been a dancer all her life, trained in ballet and modern dance, and knew dance as an activity in which visual observation is critical for performance. Many times an instructor will ask students to reenact a complex movement or sequence they had just been shown. Some students, Cross noticed, were far more adept at picking it up than she. Did their mirror neurons simply work better than hers? Other times, movements could be frustratingly obdurate for everyone. In the fall of 2006, Cross' troupe, the Dartmouth Dance Ensemble, was gearing up for a performance of Laura Dean's 1982 composition *Sky Light*, a challenging and unusual postmodern work: gold satin, heavy spotlights, geometric shapes. Cross struggled to grasp it. So did her teammates. *What happened to our mirror neurons?* Cross wondered. The flurried arrangement of Dean's stomps and spins weighed on Cross' mind in the lab where she worked during the day. She wanted to know the parts of the brain that might be responsible for translating the movements that the dancers saw into movements the dancers did. One day, she brought it up with her mentor, the lab's director, Scott Grafton. "He said, 'You should study that,'" Cross recalled. "I said, 'Wait, really?'"

A short time later, the dance company (eight women, two men) gathered at the lab for six weeks of experimentation. In an fMRI scanner, each was shown short clips of the new dance they were learning from *Sky Light* interspersed with clips from other, nonrehearsed but similar dance movements. As they watched, the dancers were also asked to imagine making the movements they were seeing. After each clip, the dancers were then asked to rate their ability to perform each movement. Over time, Cross, along with Grafton and Antonia Hamilton, the coauthors on the study, defined patterns of different brain activity as the dancers watched the clips they had never seen versus clips of sequences they had seen before and perhaps had even performed themselves. When the dancers observed movements they had rehearsed, the brain activated in different regions than when they observed new movements. When they came upon movements they knew they could actually perform with a high level of expertise, the brain was even more active. They took it as confirmation of an idea called "action simulation"—the way in which we might internally simulate an observed movement using similar brain regions to the ones used to execute the movement ourselves. The fact that these were experts showing the results was additionally promising. "We got this really focused, neurally efficient response," Cross says. "The better you are, the more you simulate in these core regions." The core regions were brain areas known to be involved in motor production. Lying supine in an fMRI scanner, however, these dancers did not actually dance. Rather, their brains seemed to be simulating the movement as they watched and imagined.

Stronger activations occurring for movements they had executed before established what the authors considered to be "a role of physical embodiment in action simulation." Experience could not only shape performance. It could shape perception, changing how we observe and plan the actions we intend to make. There was a loop between how we watch something and how we perform it, and that loop fed both forward and backward. It informs an expert basketball player that the fingertips will be the giveaway to a missed free throw, not just because he knows what to look for but also because he has felt that sensation before. He is an expert in that sensation. Prior physical experience enhanced and focused the activity in areas involved in both action observation and imagination. "That really showed embodied expertise for the first time," Cross said.

The brain areas that Cross recognized in embodied expertise—specifically the inferior parietal lobule (IPL), the superior temporal sulcus (STS) and the ventral premotor cortex (PMV)—are now accepted as focal points of the action observation network.

The AON is somewhat misleading: Action may be observed, but often never reproduced. We take in what we see—say, a person jogging by or a baby crawling—and we simply absorb it. The stimulus

EMBODIED EXPERTISE

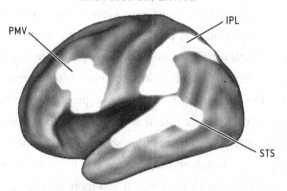

Ventral Premotor Cortex (PMV): The brain region originally identified to contain "mirror neurons" in macaque monkeys. In humans, the PMV is said to be a crucial area for translating an observed action into a desired motor act.

Inferior Parietal Lobule (IPL): Residing at a juncture between the visual, auditory and somatosensory regions, the IPL is best known for its role in integrating signals from different modalities (such as visual, auditory and motor signals).

Superior Temporal Sulcus (STS): Often implicated in "social" cognition, this brain region integrates signals from different sensory modalities. Posterior portions of the STS are particularly crucial in high-level motion processing, extracting meaning from spoken and unspoken social exchanges, and perceiving facial movements and eye-gaze (where others are looking).

Brain regions are more active when watching dance movements the observer has practiced versus ones she has not. Cross describes this as "embodied expertise."

rattles around certain brain regions, regions that are implicated in action, but we do not act ourselves. Some believe, as Sperry did, that we are getting ourselves ready in case we have to. We can make sense of what we are seeing because we have done the same actions before. We can imagine what it would feel like to do them again. We can even simulate the way it would be done. This happens implicitly and subconsciously. It forms a hazy template in our brain for the actions that we can draw from again and again, so we don't have to always embody what it feels like to reach out and shake an extended hand. We just do it anyway.

This template is known as motor resonance. A visual representation is intrinsically stamped onto a motor representation. It is a form of recognition that provides the basis for the prior knowledge that enables our Bayesian modeling capability to function. The better I understand that template, the more often I have seen the action before, the easier it is for me to reproduce the movement. And the better I can see myself as the actor, the better that image should resonate. The AON has been shown to be significantly more engaged when I watch another human bite food off a spoon than when I watch a dog do the same. It was more engaged in dancers when watching a move that was performed before than when shown a move they had never performed. It is a kind of bias toward actions I know I can do myself, as well as actions that are being done the way I might do them. The observed action matches up with what I recognize from my own movements. This might explain why the popularity of an athlete like Stephen Curry (John Krakauer's favorite player) could transcend the hype for a typical NBA star. His style—shooting jump shots from long distances, playing below the rim, chomping on a mouth guard—is not unlike that of millions of recreational players, and so, in Curry, we cannot help but see a bit of ourselves. A motor template for how we would want to play basketball is formed.

We also encounter actions performed by those who do not look or seem like us at all. There are actions we cannot reproduce, and yet we can still make sense of them. Perception of movement is deeply engrained in us. Walt Disney found that he could test animation applicants by asking them to transform a half-filled sack of flour into a lifelike cartoon. The best animators created sacks that moved in ways that we would instantly anthropomorphize and become attached to with our feelings. The AON ingests that information and prepares the motor areas for the possibility of movement in response. "It could be the case that when you learn from vision— just sit and watch someone do an action I'm supposed to reproduce later—this is all happening in the back of the visual cortex," Cross says. And then when you do have to reproduce the movement, the brain switches to the motor regions. But no. The regions are integrated. Seeing primes the body for doing. It can be as specific as you want it to be. If I'm watching Serena Williams gently curve her wrist in preparation for a serve, the part of the motor cortex that controls my own wrist gets excited. If my focus changes to her shoulder, then that excitability shifts to the area for the control of my shoulder, and it diminishes in the wrist. The AON is like a refrigerator always humming in the background. In 2009, Cross did a study in which she gave no explicit instructions to the participants about whether they actually had to learn the dance sequences they were asked to watch. Yet the AON was still engaged, preparing the motor representations anyway. The brain was soaking up the visual information even if the intention was never to actually put it into action. The brain was ready just in case. Furthermore, it is not only wired to take visual cues and turn them into movements immediately. It is equally sensitive to information it might not need right away, but could rely on later, just like the violinists who were surprised to learn they could finger the strings with their right hand as well as their left. *The brain is not a sieve; it is a sponge.*

Cross is about five foot three with the trim, immutable frame of an endurance athlete who has never considered taking the elevator. She is ebullient and lighthearted, with a somewhat rhythmic quality to her speaking. She has ombré hair transitioning from dark chestnut to fire-engine red where it falls on her shoulders. Her laboratory is now at Bangor University in northwest Wales. She has a large office with a partial view of where the swift and narrow Menai Strait empties into the Irish Sea. This is mining territory, not the pastoral sheep-and-black-cattle farmlands the Welsh countryside is known for. Bangor was founded by the quarrymen who wanted their sons to be educated off their earnings from the heather-blue slate they sold to make the roofs of Buckingham Palace, King's Cross, the Welsh Assembly and Harvard University. The sills outside Cross' office in the Brigantia Building, not surprisingly, are made from slate. Her shelves are lined, in the true mode of an aesthete, with books on dance, art and psychology. A tiny silhouette of Michael Jackson is pinned to her bulletin board. Michael Peters, Jackson's choreographer for the music videos "Beat It" and "Thriller," said once that the King of Pop's knack for picking up dance moves was unlike that of anyone he had ever encountered. "Here are these people that have spent thousands of dollars training to be dancers all their lives," Peters said. "And this kid walks into a room, and you say this is the beat, and he does it." Cross always wanted to know how.

She believes the answer exists somewhere in the AON, which was borne out of the hazy mirror-neuron fracas but offers a more definitive, albeit much larger region of neural acreage implicated in perception-action activity. Few dispute the existence of the AON, although there are debates about its role and function. Human mirror neurons, for all we know, could be in the AON somewhere, quietly parroting the actions swirling around us all day. The sites where they are thought to reside—namely, the ventral premotor cortex and inferior parietal lobule—are representatives of the AON. But a kind of stigma has tainted most research on actual "mirror" properties today. It draws a sigh from Cross. "I don't think the media frenzy with it started off in a bad way," she says. "But then it kind of went bananas. That wasn't just the media, let's be honest. There was some pretty questionable research done by a few groups trying to extend mirror neurons for every disorder." Cross is careful to point out that the original findings, and all that is really conclusive still, showed *similar* but not identical neural responses to perception and action in the macaques. But their very existence still eludes our grasp. "Have we actually recorded in those regions of the human brain? No. Not yet."

I learned you have to be careful about how you reference "mirror neurons" today. Antonia Hamilton, who works on social interaction at University College London and worked in Grafton's lab with Cross, said the term could still be a red flag for some. Paul

Gribble, who studies the perception-action relationship at the University of Western Ontario, gently admonished me when I brought it up. "We try not to use that phrase," he said.

As a dancer her entire life, Cross was constantly surrounded by performers who seemed better than her at absorbing and reenacting sequences that the teacher had demonstrated. This is what we know of as observational learning, and not everyone is equally good at it. The child psychologist Albert Bandura was the first to popularly distinguish between modeling and imitation. He wrote, in 1977, that "virtually all learning phenomena resulting from direct experiences can occur on a vicarious basis through observation of other people's behavior and its consequences for them." In his famous Bobo doll experiment, he had children watch footage of an adult kicking, hurling and pummeling an inflatable bop bag with a mallet. The children, presented with the same doll, reacted just as aggressively. Force begat force. The children did not simply imitate the adults; they embellished their own techniques of assault. For Cross, observational learning in dance classes became a source of frustration. "I sometimes think, 'How are they translating that visual information so well into their own bodies?'" she said. "'What is it that they are doing? Are they making micro-movements while they're watching?'"

Now she believes it has something to do with their embodied expertise, but it took some time to get there. In 2004, Beatriz Calvo-Merino at University College London recruited expert ballet and capoeira dancers into the neuroimaging scanner with an intriguing idea. She showed each of them 24 different three-second videos of various capoeira and ballet movements, and asked them to press a button indicating how tiring they thought the move was. She and her collaborators found that when the ballet dancers watched ballet, or when the capoeira dancers watched capoeira, their neural activity was greater across the left dorsal and ventral premotor cortices, and the posterior superior temporal sulci and bilateral intraparietal sulci. These were landmarks of the AON. There was nothing like that when they watched movements they were not familiar with, even though many of the motions—jumping, swinging arms and so forth—were aesthetically similar.

The researchers surmised that those areas of the AON must be particularly sensitive to learned action patterns and movements. To test just how sensitive the AON is, Calvo-Merino followed up with 24 male and female professional ballet dancers and showed them video clips of movements made by their own sex, as well as moves performed by the opposite sex. From the scanner results, they could dissociate brain responses related to motor representation from those related to visual knowledge. All of the dancers had visual knowledge of every move—they had been around male and female dancers their entire careers. But *motoric* knowledge was different. They had acted out only the movements specific to their own sex. Using the scanner results, Calvo-Merino could discern greater activation in premotor, parietal and cerebellar regions when the dancers viewed moves from their own repertoire, compared to the moves they saw often but had not actually performed. The AON, in other words, could distinguish between observations it knew from experience and those it just knew from seeing. In fact, physical knowledge was a major influencer of neural activity.*

---

* Combined, this paper and Emily Cross' study on embodied expertise have been cited more than 2,300 times.

"If you think about it, it is kind of amazing anything like this can happen at all," Gribble said. "If I'm watching you move your arm, the information I'm getting is a series of changes in light on my retina. Relating that information on my retina to my own representations of my arm—that's a very distant mapping." But Gribble has also seen evidence of engagement in the somatosensory cortex as people watch others trying to perform a motor task. That part of the brain is normally thought of for its role in touch, or sensing where the body is in space. But there it is, perky and active, somehow involving itself in the transfer of vision to action, even in the absence of any overt motion. "Information can be represented in multiple domains," he said. "And when these different regions talk to each other, some kind of translation occurs." He added, "Maybe somatosensory areas are more of a bridge—a stepping-stone between visual and motor." Whereas Gribble has been focused on the somatosensory area, Hamilton has pinpointed the prefrontal cortex as her region of interest. That is because, in her studies on autism, she has found that people with the disorder do not reflect the same level of engagement in that area when they have to decide to copy a motion or not. "Imitation is very selective," Hamilton said. "People are all the time making judgments about, 'Should I copy that or not copy that?' What we find is that kids with autism, if you tell them explicitly 'Copy me now' they can do it just fine. But what they're not doing is deciding spontaneously on their own when they're going to copy." In this case, the decision-making area of the cortex may not be translating the information it needs to be.

When I arrived at Cross' lab, her students had just completed an exhaustive study involving 15 adolescents aged 12 to 16 and 18 young adults aged 18 to 30. The goal was to understand how the brain might learn to encode different actions, like ones it has performed, ones it has seen and ones it has never seen at all, at different developmental stages.

I asked Dilini Sumanapala, the PhD student conducting the study, to reenact what took place. And again, this being a neuroscience lab, I expected assorted rigs, scanners, caged animals, fume hoods, and at least a whiff of formalin. But I'm learning. Instead she cued up a pulsing hip-hop track called "Like a G6" by Far East Movement. Just in case there was any trepidation about dancing in public, the lab room was fitted with venetian blinds. Sumanapala collected the biometric data using a Microsoft Kinect system right out of the box. The Kinect does not require a controller; it records and responds to your body movements via a motion-sensor camera that produces a real-time silhouette of your actions on the television screen. You can track your own movements as you step in time to the avatar on-screen. In this case, the avatar was Angel, wearing a popped collar on his navy coat and gray sweatpants rolled up at the ankles—the "least busy outfit" in the game, Sumanapala said. The participants danced with Angel for more than an hour every day for five days.

That amount of practice should be enough to begin the formation of motor resonance. When the researchers put the participants in the scanner, all they needed was to show them the recorded footage of their actual dancing with Angel to trigger patterns of activation in the AON. The patterns, as described by a complicated analysis procedure, are then compared with those they gathered when the participants were shown footage of a randomized dance sequence.

A third piece of footage was not visual at all; they simply played a song through the scanner headsets. What the investigation revealed was that the AON seemed to be capable of discriminating between the various inputs: audio, visual, motor.

"This is a stepping-stone toward us finding different types of patterns that are specifically associated with different forms of experience," she said. "If you're a ballet dancer and you've trained for 10 or 15 years, you've trained *Swan Lake* and you know that entire routine as Odette, is your motoric representation of *Swan Lake* the same as someone else's?"

"Probably not," I said.

"Then we can get at questions about how we all encode experience," she said. "If we can start to pin that down even further, we could potentially think of putting that information into domains like prosthetics. Essentially, we [might be able to build] a common code: [If we have] this person who wants to perform a grasping movement—how do we encode that? How do we put [that pattern of activation] in their brain so they can perform this movement?"

I was reminded of an experiment from Gribble's lab in Ontario. He had worked with four groups of participants, each assigned to learn how to control their reach toward various targets at a table similar to the one I used with Aaron Wong at Johns Hopkins. But each reach, made while gripping a robotic arm, gets perturbed by a slightly imperceptible force on the robot to make the task more difficult. This is called a "force-field adaptation" task and is one of the most elementary experimental paradigms in motor research, because it addresses how we learn. Participants had to learn to adjust their trajectories in order to reach the targets. The first group, a control, had no practice. But the second group got to watch a re-play from a video camera placed above. They had to count the num-ber of targets the first group reached. Then, when it was their turn to try it, they proved to be dramatically better, decreasing the learn-ing curve, Gribble said, by as much as 20 percent.

This should not come as a huge surprise. A batter who watches film of a pitcher will typically fare better than one who chooses not to, simply because he knows more about what to expect. But when Gribble took the third group, he added a distraction as they watched the replay. They had to count arbitrary numbers on top of counting the targets. A fourth group had to draw figure eights. Interestingly, it was that latter group that fared worse than everyone, other than the controls. By occupying their motor system with something else, their brain could not absorb what they were seeing as well as the rest. "If what's happening when we watch is our motor areas are activating in some way, if we interfere with that activation by hav-ing you actively do something else, that should reduce the benefit," according to Gribble's hypothesis going in. The results confirmed it. "This was not an explicit strategizing," he said. "It was really an implicit activation of the system."

Still, he is quite certain there are those of us, naturally, who are strong physical learners; we need to handle the paintbrush in our hand in order to make the movements necessary to copy a piece of art. Others can more adroitly follow along as a coach draws the *X*s and *O*s of a play call on the whiteboard. Still others need to hear the instructions first in order for them to piece the puzzle together in their minds. Why there are such differences has to do with how the AON receives and processes the information. In general, it is believed—based on some empirical evidence—that a linear relationship exists between modes of sensory input and AON engagement. That engagement with visual experience is said to be stronger than auditory experience; audiovisual is stronger than visual; physical is stronger than audiovisual; and visuomotor is stronger than physical.

The more we layer our experience across different sensory modalities, the greater the AON response seems to be. It is not enough to watch an expert pianist on the keyboard to learn how we ourselves should play. We need to feel the keys on our own fingertips. "It's about having more channels available to you," Cross says. "The more channels through which you experience an action, the better

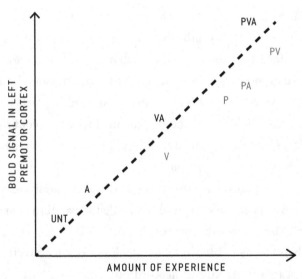

Hypothesized relationship between activity in the left premotor area and different sensory modalities. Starting with untrained (UNT) in the lower left-hand corner, AON activation should increase as you add audio (A), visual (V), visual + audio (VA), physical (P), physical + audio (PA), physical + visual (PV), and, finally, physical + visual + audio (PVA). The more channels there are available to you, the more engagement there is of your AON.

the learning is and the more engagement we see in the AON." That is not to say that we learn only by copying what we see; blind people are counterexamples of that. But simply observing is hardly worthless, and in fact there is increasing evidence—mirror neurons aside—that how we watch something will go a long way toward how we ultimately perform it. "The processes between doing and observing are shared," said Dace Apšvalka, another PhD student in Cross' lab who has been examining the motor representations of observed actions. "It's as highly specialized as if I needed to do it myself."

Even if it is shared, though, "It may be the case that not everyone is able to robustly encode everything they're observing," Sumanapala told me. Some might be capable of encoding better than others. There are important implications to this research, which is not just for dancers attempting to learn a new position or football teams hoping to make halftime adjustments, but also for stroke recovery and physical therapy. Patients with weakened limbs allow their bodies to be stretched, pulled and contorted by therapists in a daily effort to regain some strength or sensation, even as the likelihood of improvement drops off radically after the first few months, as described by John Krakauer's recovery window.

In some of these patients, though, the rigorous physical exercise might not be reverberating in the neural networks responsible for reproducing those movements as well as assumed. "They might not be the best candidates for a rehabilitation procedure that involves regaining motor functioning," Sumanapala says. Others might benefit from virtual therapy on top of the physical therapy. If such strong connections exist between the motor system and the visual system, who is to say you cannot access a debilitated motor system through the visual system? "You can imagine giving them some visual regimen that they could do at home"—like a highly specific cognitive game—"such that it primes the neuroplasticity in their sensorimotor system," Gribble said. "So that will be primed when they go to the clinic."

The increased AON engagement when viewing actions we ourselves have done before, and have done often, could essentially be described as a handoff from seeing something to *knowing* it. It resonates better with us, and from this we can extrapolate certain conclusions as we watch other people. In the early 2000s, Rizzolatti wrote of a phenomenon called "action understanding," in which a passive observer can produce an inference about the goals and intentions of another person simply by watching her movements. "The proposed mechanism is rather simple," Rizzolatti wrote. "Each time an individual sees an action done by another individual, neurons that represent that action are activated in the observer's premotor cortex." Many believe this inference is generated by our own motor systems. The AON, as it were, is also Bayesian. When we see a person across the street raise his arm in the air, he could be either hailing a taxi or swatting away a wasp. But we can estimate his intention by a contextual cue, perhaps the expression on his face. This becomes the prior in the Bayesian paradigm. The prior in this case is a prediction about what someone else will do. But with that prior, we can estimate a goal—if his face is distorted in panic, we can surmise it is because of the wasp. From the goal, we can predict the motor command. From the motor command, we can predict the kinematics required to fulfill it, based on our *own* kinematic experience. The kinematics we see from someone else can be compared to the ones we have, generating a prediction error. This error is then used to update the inferred goal.

That sounds complicated. But let's say you are invited to watch identical movements made by two men, Dr. Jekyll and Mr. Hyde. They are both grasping a scalpel and slicing into a human patient. However, Dr. Jekyll's intention is to aid the patient; Mr. Hyde wants to injure him. How can we infer what each man's intentions are? The Bayesian instincts of our AONs might account for the setting of each procedure—if we see the action taking place in an operating theatre (and not some basement torture chamber), there would be a large prediction error for the intention to "hurt" compared with the error for the intention to "cure." The AON is capable of inferring a unique intention even if faced with identical movements.*

Reading these sorts of subtle cues is critical not just for baseball hitters trying to guess whether the incoming pitch is a fastball or a curveball, but for many kinds of ordinary social interaction. In fact, it is what keeps us on the dance floor. When you are trying to keep up with a dancing partner, the task is to make your body do something that the actions you see might *feel* like, based on how you predict they would feel. It prepares an internal description of the action and, with it, an ability to know how to respond in the future. "The brain is making its own predictions nonstop," Cross says. "If we just reacted, we'd be dead."

In 2015, Jason, Jordan and a handful of collaborators, including Paul Sajda, put Columbia baseball players back in the fMRI. They wore the EEG caps again and tapped a finger to respond to the pitch

---

* The Dr. Jekyll and Mr. Hyde thought experiment comes originally from Pierre Jacob and Marc Jeannerod, who used it initially as a critique of the predictive coding suggestion. James Kilner, Karl Friston and Chris Frith, however, reinterpreted the parable in the way I have used here.

simulation. They responded to 450 pitches. Then, when the simulation was over, they remained in the bore of the scanner. They were instructed to "focus on the cross on the screen." They remained there, lying supine, unmoving, for five minutes.

In another room, Jordan prepared what is called a "resting-state scan": an analysis of the brain in meditative quietude, devoid of external provocation, humming softly to its own autonomic demands. It is a moment of tantalizing clarity, the dust settling after a violent windstorm. Researchers believe these may be critical periods of learning, when the brain takes stock of what it has just been trained to do and begins to forge it into memory. The brain reshapes and reorganizes itself by bolstering the connections required to more easily and efficiently produce specific commands to the muscles. Hitting a baseball is not a problem the brain can think its way through; it requires those connections to be extraordinarily sound. Jason and Jordan wanted to see where those connections might already be established and if there were differences between the experts and the novices. They believed that 450 pitches on a difficult simulation would be enough to produce some hypothesized learning effects even within hitting experts.

They focused on the SMA as the "seed region," the kernel that might sprout connections to other regions. This was a functional connectivity analysis, meaning that they were looking for neural relationships between regions that were not necessarily anatomically tangential: long-distance neural relationships. Sure enough, their DTI (diffusion tensor imaging) assessment highlighted increased connectivity between the SMA and the left insula, an area buried deep within a cavernous sulcus. It is not close to the motor cortex. It is not usually implicated in motor studies. It is a somatosensory region, known as a switchboard for the afferent signals about the body's homeostasis, such as whether the periphery senses pain. But in 2005, a study of musicians found increased activity in the left insula when the participants listened to a piece that they had played previously versus one they had never heard before. It was embodied expertise. Additional studies have postulated that the insula plays a role in error processing, guiding us toward better outcomes and helping us learn from our mistakes. Just by virtue of their motor experience, expert hitters might have established a connection between two disparate brain regions that reinforces their ability to learn from errors. Jason was so excited by the results he kept pictures of the slides on his phone. "It's the areas involved with object recognition, decision-making, things like that, that are connected with the motor area," he said. "And it's particularly strong in players."

In some cases, performance can improve without needing to move or watch anything at all. "Motor imagery" is the mental rehearsal of actions without the actual movement itself and is a heavily relied-upon technique of sports psychologists, increasingly growing in popularity. It has occasionally been related to meditation. In both meditation and motor imagery, some physical effects, such as vasodilation, are not only imagined but also experienced. Phil Jackson, the legendary coach of the Chicago Bulls and Los Angeles Lakers, earned his nickname, the Zen Master, for his attraction to meditation and Zen as a routine to help his players relax and remove themselves from the burdens of pressure. But, more and more, trainers are urging athletes not to withdraw from visualizing difficult athletic scenarios but to embrace them. Imagination can be a powerful tool. Just as an observed action can trigger our motor system into attention, there are indications that merely an envisioned action involves similar motor-related areas, including regions involved in motor execution. Savvy performance coaches have taken that as support for the notion that you can practice a skill just by mentally rehearsing it. The most famous example is Michael Phelps, who has been known to use blacked-out glasses to better focus on the conjured race unfolding in his mind. And ahead of the 2014 Sochi Games, American aerialist Emily Cook told *The New York Times* that she wrote entire scripts of what a perfect jump should look and feel like, and she would read the script into a recorder. She would then play the recording with her eyes closed. "I was going through every little step of how I wanted that jump to turn out," she told the newspaper, before finishing eighth, a career best.

The psycho-neuro-muscular tripartite relationship is not just about confidence. There is empirical support that just mentally simulating a movement can activate brain areas like the SMA, dorsal premotor cortex, supramarginal gyrus and superior parietal lobe—the common regions responsible for preparing and producing the movement. It has contributed to a hypothesis of "functional equivalence": that cognitive processes seem to share, to some degree, representations, neural structures and mechanisms with the *seeing* and *doing* aspects of the motor system. Two studies found that simple visualization techniques could assist golf putters, including as a cure for the yips. Another looked at imagery among skilled tennis players benefiting their ability to return a serve. There has also been growing support for the potential advantages of mental imagery *during* action observation, or imagining yourself going through an action as you watch it, such as the Dartmouth dancers did. Various studies have shown that action observation plus motor imagery (AO + MI) increased activity in motor areas compared to observation alone.

Of course, there is reason to believe that some degree of AO and MI is happening concurrently all the time as we move throughout our daily lives, observing actions in the world and silently rehearsing them, as we might do them ourselves. If that capacity did not exist within us, we might not be here. It is difficult to imagine how we could have survived if our actions did not align with the consequences we intended, based on the experiences we have seen or enacted before. "We didn't evolve just by inventing the wheel and everyone having to reinvent it themselves," Cross said. "We watch other people and we say, 'Right, OK, now I'm going to do that.'"

That experience might shape the prism through which movement is understood, like a pair of contact lenses that habitually sharpens over time, is no more outlandish to Cross than how fluency in a language will help one read a foreign menu. It acts as a key to unlocking more details, more clues, more context. The basketball player who has been to the free-throw line 1,000 times can thus know more implicitly the subtle differences between a make and a miss. The dancer who has made the motions before can thus better anticipate what should come next. The brain inside the driver inside the race car is extraordinarily engaged with the road it has ridden and will ride again.

"If I listen to a piece of music, I'll hear the overall melody," Spackman told me. "I'd often do this with race drivers—I'll play them a tune. I'd say, 'Now, hum the tune back and tell me what you heard.' I'll play it back again and I'll say, 'I want you to concentrate on the oboe coming in and out,' or 'Here's the clarinet on the top.' And they'll say, 'I didn't hear any of that.' Obviously all of that made the melody. All of those details made the music sound what it was. But I wasn't aware of all the detail. Now, someone like Mozart had the ear to hear all these things and be completely aware of them all on one go.

"The same analogy could be used in racing," he said. "The good ones hear all the elements that make up the symphony."

8.

THE BODY IN SPACE

HOW TOM BRADY WON SUPER BOWL LI

I n 1900, Charles Sherrington, 43 years old and frustrated by the slow progress of his taxonomy of the sensory and motor nerves, had about had enough. "The paths of sensory conduction in the cord are in a fearful mess," he wrote in a letter to Theodore Acland, a friend and physician at St. Thomas' Hospital. Sherrington was rightfully exhausted. He conducted all his experiments himself, dissecting, preparing, staining and sectioning nerves with his own hands, then personally counting the fibers. He kept no secretary, preferring to maintain his own personal correspondences (of which there were many) by mail. He often took long walks and, upon returning inspired, would manically sketch diagrams on the walls of his office with a pencil. He had become consumed by the quest to understand the details of the nervous system by individually severing every root from the lumbar-sacral region of the spinal cord and analyzing the behavioral consequences in a variety of animals.

His preoccupation with the fine details did not eclipse his over-arching desire to understand the body as a whole, and he took his influences from a wide range of sources, which could often side-track him. His former student John Eccles recalled a morning when Sherrington came into the laboratory jabbering excitedly about a cat he had seen on his walk to work. "He had seen a cat walking solemnly on a stone wall that was interrupted by an open gate," Eccles wrote. "The cat paused, inspected the gap, then leaped exactly to the right distance, landing with ease and grace and resumed its solemn progression. A very ordinary happening, yet to Sherrington on that morning it was replete with problems for future research. How had the visual image of the gap been transmuted by 'judgment' into the exactly organized motor mechanism of the leap? How had the strength of the muscle contractions been calculated so that the leap was exactly right for the gap? How had the motor machinary [*sic*] been organized so that there was this elegant landing on the far side of the gap? How after the landing was it arranged that the stately walk was resumed?"

His nerve mapping had helped Sherrington distinguish several paths of afferent input toward the spinal cord. There was a path for touch, obviously, with long fibers plunging directly into the gray matter of the dorsal horns. There also appeared to be tracts of fibers producing sensations of pain, warmth and cold. And there was a fifth, even larger, path, whose root fibers swept right up to the up-permost tip of the cord. But when he lesioned the anterior roots, an operation that should have killed off the motor nerves, nearly half the fibers in the spinal nerve were unaffected. Sherrington con-cluded that these must be sensory fibers, not from the skin or joints, but from the muscles themselves. This seemed odd—the muscles, of course, did not appear to contain any type of sensory organ. In his letter to Acland, Sherrington referred to the path as the one responsible for "muscular sense," which underlies the "unconscious processes [to] regulate [and] help to coordinate movements—especially such habitual reactions as walking, posture, etc."

It was not the first time Sherrington had tried to examine the amorphous notion of "muscular sense." Many others had tried as well. In 1798, E. B. de Condillac, a French philosopher, observed that our sense of the weight of our limbs and the resistance we feel as we stretch our muscles might be something he called "active touch." The Germans called it *Muskelsinn*. Charles Bell listed three qualities of his version of "muscle sense": pain and fatigue, weight and resistance, and movement and position. "Muscular sense," David Ferrier wrote in *The Functions of the Brain*, "is applied to the assemblage of centripetal impressions generated by the act of muscular contraction in the muscles themselves, as well as in the skin, fasciae, ligaments, and joints." One popular notion of the nineteenth century was that muscle sense had to do with the signal being sent from the brain to the limb, giving off an abstract feeling of "effort." That view was dismissed as soon as Sherrington demonstrated that the brain is not necessary for many movements at all.

But what was it, exactly, that gave the impression of "feeling" from within the limbs themselves? With our eyes closed, we can still maintain our balance, reach for coins in our pocket or touch our fingertip to our nose. "A man waked suddenly from sleep in the dark is aware of how he finds himself lying," William James once wrote. We know the positions of our body as we move about, enabling us to avoid obstacles we encounter or reach for objects at a distance. Our muscles can seem heavy and sluggish after rigorous exercise. We can drive a car while looking at the road and apply whatever subtle contractions are needed to speed up or brake. And more recent work has been devoted to understanding how this knowledge—or "proprioception," a term that Sherrington coined—actually informs our actions, rather than the other way around. In 2015, researchers in Shanghai, Sydney and Canberra tested elite athletes such as gymnasts, swimmers, dancers and soccer players and found that the proprioceptive acuity they reported from their ankle, shoulder and spine accounted for 30 percent of the variance in competition level achieved (while also factoring in years of sport-specific training), suggesting that similar proprioceptive tests might be useful in talent identification. "This proprioception," a patient once told Oliver Sacks, "is like the eyes of the body, the way the body sees itself."

Sherrington addressed it mainly by recognizing that a simple motion like walking might be automatic, but it is not stereotyped. There is a constant stream of input regulating the motion, even if it is not coming from the cortex or the mechanoreceptors of the skin. Muscles, he deemed, are not merely passive instruments moving at the whim of the nervous system. In 1898, he posited that they might be in possession of receptors of their own, an additional method of sensation, like a sixth sense. This meant that they could "have some voice in their own conditions of service, perhaps ring themselves up and ring themselves off." He termed it muscle "receptivity" and credited it with addressing "the taxis of execution, the management—from rough adjustment onward to minute refined finesse—of the acts of our skeletal muscles." In *The Integrative Action of the Nervous System*, he had a new term for this sixth sense. He called it "proprioception," which he formed by combining the Latin word *proprius*, meaning one's own, and "perception." He defined it as "the perception of joint and body movement as well as position of the body, or body segments, in space" and distinguished between proprioception, exteroception (the sense of the outside world's interaction with our body, mainly through touch) and interoception (the sense of the body's internal state).

Today we know, as Sherrington identified, that proprioception is informed by what are called "muscle spindles," while information about the force we are exerting is provided by Golgi tendon organs (discovered by Camillo Golgi). Combined with the skin's mechanoreceptors, this system gives us the intrinsic information we need so we can estimate changes in body position. But some don't look at proprioception the same way as the other five senses, such as hearing, which relies upon sound waves stimulating the inner eardrum. Muscle sense is not just a recipient of signals; it may also be shaped by memory and bolstered with learning. "In this understanding," writes Jia Han in a 2016 review, "proprioception can be defined as an individual's *ability* to integrate the sensory signals from mechanoreceptors to thereby determine body segment positions and movements in space. In other words, proprioception is not merely a physiological property, but rather, it has both physiological (hardware) and psychological (software) aspects."

That ability, Han discovered, might be more responsible for skilled motor performance than we previously realized. In a study two years earlier, he and others gathered 100 elite Chinese athletes (gymnasts, swimmers, dancers, and badminton and soccer players) and assessed their proprioceptive acuity at five distinct body sites: ankle, knee, spine, shoulder and hand. Their results were better across the board, in all five joints, than those of nonathletes. This keen sense of oneself could be the result of the hours of physical tuning the athletes devoted to their sport, or it could be the reason for it. A study in 2016 by researchers in Vancouver asked 28 participants to walk on a treadmill and match the foot height during the swing phase to the height of a virtual obstacle shown on a screen in front of them. Before the task, the participants were assessed for proprioceptive acuity. Lo and behold, those participants who showed better acuity before the task wound up improving more at the obstacle-avoidance task than their fellow participants.

The role of body awareness to motor skills is more acutely observed in those rare cases when the two are involuntarily divorced. There was the young patient of Dr. Adolph Strümpell—as the story is told by James in *The Principles of Psychology*—who had no means of sensation other than his right eye and left ear. They could cover the boy's eyes and carry him around the room, lay him on a table, contort his arms and legs into unsightly postures, and the boy would have not a clue. When the eye bandage was removed, the boy would be astonished. And, more recently, there is the case of Ian Waterman, the subject of a documentary and a play based on Jonathan Cole's 1991 book, *Pride and a Daily Marathon*, who was found to have a nervous condition that attacked his sense of touch and his sense of movement or positioning below the neck. His perception of temperature and pain remained intact. But Waterman could no longer tell where his body was positioned unless he could see it. And the lack of feedback from the limbs irrevocably impaired his body's ability to produce controlled movements, to the extent that he became wheelchair bound and effectively paralyzed. During rehabilitation, Waterman relates a funny story about sitting next to an attending nurse, when his arm accidentally grazes the woman's chest. She slapped Waterman in the face, stunning him. "What was that for?" he asked. He had not realized where his arm had drifted nor could he get any feedback from the touch of the woman's blouse. In fact, without looking at it, he could not feel his arm's presence at all. In a 1997 BBC documentary, Waterman recalled the early days of his condition—which was triggered by a gastric flu—lying on a hospital bed he could not feel, like a quadriplegic. "What use is an active brain without mobility?" he asked, clearly wishing he, at that moment, could be more like a sea squirt and digest his own brain.*

No one had heard of a disease attacking the nervous system in such a way. Waterman was not paralyzed; his limbs were motile. But without the limbs supplying their own sensory input, the brain was essentially left in the dark. It was similar to an experiment by Sherrington in 1909 when he cut the fourth, fifth and sixth lumbar nerves of a cat and isolated its knee-extensor muscle, called the vastocrureus, which in previous experiments he had demonstrated to be critical to the muscular sense required for the knee-jerk reflex. With the afferent roots now severed, however, the knee jerk could not be elicited. There was also no trace of decerebrate rigidity. The reflex properties Sherrington had shown to be automatic, regardless of consciousness, were no longer applicable to a limb stripped of its proprioceptive acuity.

Waterman's nerves were not severed, which meant he could eventually relearn—through extensive practice and visualization techniques—to sit up. It took four months for him to be able to put

---

* Today you can plunk down $99 an hour in Brooklyn to experience what Waterman felt in so-called sensory deprivation tanks, in which you can float in an Epsom salts bath in the dark. Reportedly, members of the Golden State Warriors, including Stephen Curry, have become adherents as a way to "unplug."

a sock on again. He had to consciously think about each movement, each muscle employed and the consequences of that employment, something we normally take for granted throughout our lives every day. For feedback, Waterman relied solely upon his eyes to inform his brain what his limbs were up to—if the lights went out, he collapsed.

Cole, a neurologist, figured out that the affliction Waterman was dealing with had to do with the nerve fibers running throughout his periphery. Nerves, like telephone cables, run from the brain through the spinal cord and out toward the ends of the limbs. But in each nerve, there are smaller cables, called nerve fibers. Some of these nerve fibers are sensory and some are motor, and they come in different diameters. Sherrington discovered the size differences while at Oxford in 1928. "Instead of the expected population of nerve fibers of fairly uniform large size," John Eccles said of the revelation, "there were distinctly two populations." Sherrington presumed that the smaller fibers were actually "underdeveloped" motor fibers. In fact, they are the afferent pathways responsible for the reception of thermal signals or pain. The larger fibers innervate the muscle spindles, a bundle of muscle fibers entwined with the sensory fibers that deliver a signal whenever the muscle is stretched. It is these spindles, which can vary in sensitivity, that indicate to the brain any changes in the length of the muscle as we move about, offering a sense of one's own positioning.

The virus that infected Waterman's body is believed to have attacked his dorsal root ganglion cells, crucial communication lines that transmit the afferent information from the periphery into the spinal cord and up the chain of the central nervous system. Though certain somatosensory feeds remained intact, like sensitivity to pain and temperature, the receptors innervated by the muscle spindles must somehow have been eviscerated. That communication pathway no longer existed, and without it, the brain or spinal cord could no longer feed its own signals to the periphery.

Today we know where proprioceptive ability begins (muscle spindles) and where it ends (Brodmann Areas 2 and 3a of the primary somatosensory cortex) and the points along the pathway for transmission. The muscular sense was always known to be involved in our ability to reach out for objects in space, whether that is a catcher backhanding a pitch in the dirt or someone going for the ketchup bottle at the end of the table. But investigators are still trying to piece together what its role might be from there. When you actually grasp the ketchup bottle, the touch mechanoreceptors let you know each point of contact on the object. Somehow you also have to know how to shape your hand in order to hold it. Andrew Pruszynski, the touch researcher at the University of Western Ontario, described for me a common occurrence that forces us to rely on senses other than vision: what happens when we grab something handed to us by somebody else. It could be a newspaper passed across the table or the baton being transferred between two Olympic sprinters. For the taker, the solution is simple—as soon as the skin makes its contact with the object, the signal to *grip* is initiated. This transfer could even be signaled by edge orientation neurons on the fingertips without the brain's involvement. But what about the giver? How does he know exactly when to let go?

"We're really interested in getting at, 'What information do you as the taker transmit to me, the giver, that you have the object in your grasp?'" Pruszynski told me. His theory is that the taker is actually subtly *lifting* the object out of the giver's hand, similar to what you might do to grab the baton off a table. That upward rise in the force is a delicate signal to the muscle spindles—your body positioning has changed. You can release. It is such a fine signal that it is practically imperceptible to vision. But the brain recognizes it. Touch, motor and proprioceptive inputs are all integrated together, forming a coordinated body that knows what it is grasping and when to let go.

February 5, 2017. Super Bowl Sunday. Yes. Have you heard? Tom Brady and the New England Patriots are trailing. Well, that's a bit of an understatement. Two minutes are left in the third quarter and the Atlanta Falcons lead, 28–3. ESPN had already placed New England's odds of coming back to win at about 0.2 percent. You have about the same odds of being born with six toes. But Brady has his offense at the five-yard line, 15 feet from the goal, a densely crowded, chaotic and strangely vulnerable position. The Patriots had won a Super Bowl two years earlier by intercepting a pass in this area. Now, three receivers spread out along the line of scrimmage; a fourth, the running back James White, stands next to him, ready to roam. The snap comes, and Brady looks straight ahead. Two receivers run their routes, jostle to get open. Brady looks ahead. White has snuck out to the left, into the flat, the soft underbelly of the defense. Brady's gaze is unmoved. Pass rushers are approaching; Brady has less than three seconds to throw it. Eyes flick toward White. A crisp pass. A catch, a spin, a tackle, a plow forward into the end zone. And you know the rest.

As far as Brady is concerned, the play was as routine as they get. He certainly made it appear maddeningly easy. He might even say the Patriots had practiced that look hundreds of times, and his familiarity with how it is supposed to unfold allowed him to complete the pass as if he were checking it off from a list of daily errands. But Brady's ability to gaze one way while simultaneously keeping track of the moving pieces in another direction—an essential skill for many elite athletes in sports—is an interesting problem for the brain. It is not just about having a practiced strategy in mind. Even when our eyes *are* trained on a target in a dynamic environment like a sporting competition, what characterizes that environment is exactly what makes it dynamic: There is unpredictability. Experts rely on prediction in order to handle the constraints of their task and their bodies. But the target, the receiver, might change direction or slip and fall. The defenders, whom you really cannot entirely predict, might obstruct the intended path of the ball flight. A quarterback like Brady needs to consider all these cascading scenarios while on a time crunch—large linemen are bearing down on him. The situation is easier if he could just direct his passes wherever his eyes are looking; the body prefers to be a slave to the eyes. Just think about how rarely we do anything in our daily lives without looking intently at the target. But in the NFL, that is what is known as "telegraphing the pass," and it has led to a lot of short careers for otherwise good quarterbacks.

So how does Brady do it? One researcher, Doug Crawford at York University in Toronto, has been asking the same question since the mid-1990s. Several years ago, in his lab, he set up a video screen with a cartoon simulation of two football players. One, an opposing player, was charging to tackle you. The other is a teammate standing open and awaiting your pass. The idea is to keep your eyes on that teammate as a third player, another teammate, materializes in the corner of the screen, just like James White sneaking out into the flat. As soon as you see that third man, you are supposed to reach and tap him on the screen. It should take no longer than it would to wind up for a throw. Accuracy is obviously paramount: You should be able to tap his location on the screen every time. But Crawford applies a technique called "transcranial magnetic stimulation," which sends magnetic pulses through the skull at certain areas of the cortex, momentarily disrupting (safely) the functioning of those areas. When applied to the posterior parietal cortex during the simulation, the participants could no longer successfully tap the screen where the third player was located. Their reach kept drifting back toward where their eyes were fixed: on the other teammate on the other side of the screen. Their movement was bound to the direction of their gaze.

We shift our eyes an average of four to five times every second. The motility of our visual system allows us to scan and absorb the pages of a book, the scenery of a painting, or the receivers on a football field, incorporating rapid eye shifts called "saccades." We can also train our gaze on the road as we drive using slow eye movements known as "smooth pursuits." But this system is inherently problematic. Vision is most accurate when we are still, which does not happen often. Our perspective on the world—and its diverse cornucopia of stationary and dynamic objects—is constantly being modified by our movement. The main job of the oculomotor system is to control the position of the fovea, which is the most sensitive part of the retina. The fovea is only about 1.5 millimeters in diameter. It covers only a fraction of the visual field. To examine an object, we have to move its image onto the fovea. To do that, we have to use the 12 muscles that move the eyes.

Saccades, those fast movements, transpire so quickly that we typically do not notice them. We can't notice them. If you look at yourself in a mirror, focusing on your left eye, then your right, then your left eye again, you will never see your own eyes moving. But we know they do. We can see someone else's eyes toggle in a similar fashion. Our oculomotor movements are not too *fast* to be noticed. What is happening, in fact, is that we *suppress* noticing them. Our window to the world would otherwise be like a television news report where the cameraperson is panning quickly around the scene of a burning building, trying to locate the source of the flames. If suppression did not occur, we might exist in a state of perpetual dizziness. And yet we perceive scenes as stable. The early twentieth-century psychologist Edwin Holt thought that the brain effectively anesthetized itself during a saccade. His intuition was accurate even if the method was not. It's actually worse: During a saccade, four or five times every second, you are actually effectively blinded. "One of the most striking features of consciousness is its discontinuity," the philosopher Daniel Dennett has said. It is striking "because of the *apparent* continuity of consciousness."

Indeed, consciously, we retain no knowledge of our saccadic gaps. The brain fills the gaps in for us. It does so by making a prediction about what is going to happen in those milliseconds of blur. This plan, or "efference copy," simultaneously acts like a warning signal, informing what is known as a visual map—a blueprint, so to speak, of what is in the area all around us. After the saccade is completed, the brain can reorient itself and update the map.

The map, Crawford said, most likely resides in that part of the posterior parietal cortex he was disrupting—the same region that Vernon Mountcastle studied so intently in his perception-action pursuits, hypothesizing, as far back as the 1970s, that it contains a "command apparatus" for the hand in space—along with a midbrain structure called the "superior colliculus" (the source of the efference copy itself remains unknown, though many have suggested it is produced within the cerebellum). It is based on not just what we can see but also what we can remember seeing. We rely on our visuospatial memory for many everyday tasks. If I want to pick up my mug of coffee for a sip at my desk, I often will glance briefly at the mug, then turn my gaze back to the computer, before reaching out to retrieve the mug. That reach is based on the remembered location of the mug, which is no longer where my attention is focused. Crawford hypothesized that the same mapping ability must be true as we move about. If you're strolling mindlessly down the sidewalk, you might pass a parking meter. Suddenly, you remember you forgot your phone in the car, so you turn quickly around. But instead of colliding with that parking meter, you nimbly remember to avoid it and hustle safely back to your car. How does the brain keep its eyes focused ahead while remembering what it recently passed?

It is not much different from how Tom Brady accurately locates his receiver on the other side of the field. Our active movement is supported by a process called "spatial updating," which the brain does in conjunction with the eyes and memory all throughout the day without conscious awareness. It is the brain's ability to compensate for our self-motion. Your midbrain "map" knew where the parking meter was in front of you *and* where it was in relation to your body behind you, even if you were never really looking at it to begin with. Crawford recently discovered that slow eye movements, or smooth pursuits, enable the map to get continuously updated, rather than a post hoc update like with saccades. Combined with the proprioceptive information being sent to the somatosensory cortex (conveniently located next to the posterior parietal cortex) from the muscles, there is now a clear picture of where the body is in space as we move through it. "The brain updates where things are in the visual space in real time," Crawford told me. "As soon as you want to make use of that information, it's available to go, 'Bing, OK, I know where it is relative to my eyes, I know where my eyes are, I know where my shoulder is, I can use all this information to calculate where it is relative to my shoulder to aim or reach or punch or what have you.'"

In 1943, the Scottish psychologist Kenneth Craik proposed that we carry a "small-scale model" of both the external world and the body's own possible actions around in our heads. This was different than the "little man" homunculus that internally mapped all the critical regions of our own bodies. This was an even more omniscient homunculus, serving as a scout for our future actions. "It is able to try out various alternatives, conclude which is the best of them," Craik wrote, "react to future situations before they arise, utilize the knowledge of past events in dealing with the present and future, and in every way to react in a much fuller, safer, and more competent manner to the emergencies which face it." The efference copy informs that model. By the time the copy is created, it also gets stored as a memory. This all happens before we even act.

I was puzzled by the idea of the same motor plan being considered both a predictive mechanism and a memory simultaneously.

"What the brain is really trying to do is *represent* what's happening out there in real time," Crawford said.

The biological reality is that we can't exist in real time. The nerves send signals at about 100 meters per second, which is as slow as an arrow fired out of a compound bow. At that speed, almost everything we consider to be happening *now* is already in the past. Every piece of sensory information we get and every command we make is delayed. This forces the brain to construct the scenario of the outside world in an interval that seems beyond our consciousness. Helmholtz, who was the first to notice the transmission delay in 1860, suggested that we live in a perpetual state of "unconscious inference," a notion that influenced Sigmund Freud's musings on consciousness later in the twentieth century. The philosophical subtext grew louder when, in the early 1980s, a somewhat controversial study by Benjamin Libet found that subjects reported the "urge" to move several hundred milliseconds *after* the brain had already commenced its movement initiation process. Later studies revealed that we don't become aware of an urge to act until our brain activity shifts toward the areas that actually control the prescribed movement. Consciousness of our intention is not afforded until a motor plan is in place. In other words, the decision to act comes *after* we have already begun the action. Even our awareness of initiation of a movement, Libet found, came after the brain had started actively planning the process.

The sensory feedback from the movement does not even arrive until a tenth of a second or so after the movement, which is why the sprinter gets penalized if he jumps within 100 milliseconds of the firing of the gun.* So in the absence of feedback, the brain has to

---

* Actually, a starting "gun" is no longer used at the Olympics. An electronic tone, played through speakers, is now the norm, after it was revealed that some sprinters, including four-time gold medalist Michael Johnson, were getting a later jump if they were positioned farther from the actual pistol. Essentially, the sound was reaching them too slowly.

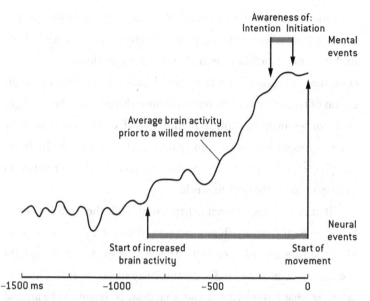

If asked to voluntarily move a finger at any time, your brain activity will begin
several hundred milliseconds before the "urge" to move the finger is even
consciously registered and up to a full second before the movement itself.
Studies on this have prompted philosophical questions concerning free will:
If brain activity can predict when a person is going to move before they even
are aware of the urge, can people be fully held responsible for their actions?

adopt different strategies to adapt. A well-known illusion is called
the "flash-lag effect," popularized by David Eagleman, a cognitive
scientist at Stanford. There are various versions, but one I like is
from Michael Bach, a vision researcher at the University of Freiburg.
A blue line is rotating clockwise around an axis, like the second
hand on a clock. As it circles, another blue line flashes at several
points along the rotation. The lines actually line up evenly, and if
you slow the rotation speed down to 1 rotation per minute you can
see that. But at 10 rotations per minute, what you *see* is the flashing
line lagging momentarily behind the rotating line.

One theory is that the brain is predicting the motion of the rotating line, which causes it to be perceived slightly ahead of the flashing line, which is essentially remaining stationary. There is a good reason for this: The brain should have a good sense of the location of things in motion, because those things could be a danger to us, or we need to capture them to survive, whereas stationary objects present less of an immediate threat.* Because of the lag in our physiological ability to perceive, we have to rely on prediction to keep up with the outside world.

That is not exactly what is happening with the flash-lag effect, however. Eagleman attributes our perception of a line flashing behind the rotating line—or in another experiment, a flashing light just behind a moving circle—to the delay in our brain's ability to *interpret* what it thinks it just saw. That delay is thought to be around 80 milliseconds. This influences the narrative you give yourself to make sense of the things unfolding around you. It is as though life as we know it is actually on a momentary tape delay, like a supposedly "live" television broadcast that runs just a few seconds behind so that producers can bleep out any errant curse words. Eagleman's contention is that the flash-lag tells us not about prediction but actually about "postdiction," an interpolation of the event after it already occurred. Indeed, in the moment the flashing line appears, we are meant to think it is not aligned with the rotating line because it is slow; it has already fallen behind. In fact, both lines are behind the present, the *now*. The reality is that our brain is playing

---

* Some humans suffer from a condition known as "akinetopsia," or motion blindness, which impairs their ability to perceive moving objects. It is thought to be caused by improper functioning of the middle temporal cortex (MT), the motion-processing area of the visual system.

catch-up every moment of every day. "As every book on stage magic will tell you," Dennett says, "the best tricks are over before the audience thinks they have begun."

If you tap your nose and your toe at the same time, you will feel the taps simultaneously. This is an illusion—the input from the nose travels a much shorter distance than that from the toe, and therefore is perceived and processed in the brain some dozens of milliseconds earlier. But the brain effectively waits for the slowest afferent impulse before informing you what it is that you felt. Consciously, we are only made aware of the taps being synchronized. "Whatever we know about reality," the cognitive psychologist Ulric Neisser once said, "has been mediated, not only by the organs of sense but by complex systems which interpret and reinterpret sensory information." We obviously grow used to this delay. Before movies became digitized, they appeared on reels of film that would be rolled through a projector. The film could not just roll continuously; in order for the picture to appear clearly and coherently, a shutter was necessary to open and shut multiple times upon each small strip, producing a subtle flickering effect. Because the shutter speed was so fast, we barely noticed it, and our brains would splice the images together without interruption. But in the early days of cinema, when the shutter speed was slower, the flickering was more apparent. An early slang term for movies was "flicks."

At other times, the delay can be more problematic. In 2002, some researchers at the University of São Paulo looked at why assistant soccer referees might make errors flagging for offside, when, in fact, replays show that the player was not out of position. This error rate was later estimated at 26.2 percent during the 2002 World Cup—an astounding number of miscalls. And the culprit, the researchers decided, might be the flash-lag effect, which makes us think that a moving object is ahead of where it actually is. Though the rule is clear—an attacker has to be *beyond* the second-to-last defender to be in violation—our own perceptive mechanism for motion is frequently fooling us.

Our predictive capacity forms the basis for a sensorimotor simulator of the body as it moves through space. It allows us to consider costs before actually committing to actions. In fact, we typically plan the extent of our movement before it is even initiated. The plan prepares the joints that will move and the torques they will require. And so it produces a template—a sensorimotor map—for the brain to rely on so it can learn how to avoid repeating mistakes. If the brain did not manage to synthesize all its inputs into a steady stream, Crawford says, our lives would be functionally upended. "Now is not just what you're seeing now, but also what you remember from a few moments ago," Crawford said. "What you've stored gets updated in a predictive fashion. For example, if I wanted to reach out and grab my scissors while I'm still looking at you, I can still remember where they are."

It is one thing for the brain to map what it sees from the couch, in a stationary environment. But we are mobile animals, encountering a landscape that constantly shifts as we move. Our brains not only need to be able to retrospectively describe what happened in the past, as *now*, but simultaneously plan for the future, as a function of what *now* could be. What Brady is seeing as he scans the field is only what already happened, but he is reacting to what is going to happen. In this light, it becomes more understandable if he mistimes a pass once in a while. "As I often point out, we can use visual feedback to guide our motion," Crawford said. "But that's just not good enough in sports or combat situations. These things have to be done in a predictive fashion. A wait of 200 milliseconds is too long, especially if things are moving. In 200 milliseconds, they're not where they were anymore."

Daniel Wolpert has long suggested that the central nervous system maintains internal representations for motor commands, too. But the task of the motor systems is the reverse of the task of the sensory systems. While sensory processing generates an internal representation of the body and the world around it based on inputs, motor processing *begins* with an internal representation of what it wants to achieve as the result of a movement, like the final page of an instruction manual. These are called "forward models" and they estimate what the world will look and feel like to me after I have moved within it. If both the current state of the body (proprioception) and the motor command are known, a prediction can be made about how the body will change as a result.

A way to know this is by recognizing how bad we are at predicting *certain* movements. Wolpert described for me a drill he calls the "waiter task." Hold a heavy object in your extended right hand— like a textbook or a tray of food. Now lift the object up with your left hand. You should notice that your right hand remained still. But if you asked someone else to lift the object out of your hand, it would involuntarily rise along with the object. It happens every time, even if you know exactly when the other person is planning to lift the object. Your brain cannot predict the external influence on your body as well as it can predict your body's movements. It then has to rely on the sensory feedback it receives, which inevitably leaves your hand momentarily trailing behind.

Fortunately, for our own movements, our predictive modeling is more reliable. If you were to hold a ketchup bottle and tap the top of it with your own finger, you can predictably generate the grip to stop it from slipping through your fingers. But if I came along and tapped the top of the bottle you are holding, even if we practiced the same routine for an hour, your grip would inevitably come several milliseconds after the tap. Every time. "I have to sense that increase in load force," James Ingram, from Cambridge, told me. "That goes up, the command goes down, to produce the increase in grip." This is the delay in action. The reason is because you cannot predict exactly when I might make my tap.

"Why would that be the case?" I asked.

"You'd think, why is the brain so stupid?" Ingram said.

It's not. There must be some reason. "We don't know why," Wolpert said. "It's interesting."

"One of the really astounding things in human motor control [is] all the tools that we use," Ingram said. "Just you getting ready this morning, all those different objects, hundreds of them, they all have different masses, different centers of mass, and you just pick one up, use it, put it back down and grab the next one. The problem that that creates for the brain is really underappreciated."

These models are what enable us to manipulate those different tools. If we relied purely on sensory feedback, we would be in trouble. "I always give the example of picking up a juice carton that you think is heavy but is actually empty," he says. What happens? Your feedback from touching the carton arrives too slowly to stop you from applying too much force and lifting the carton wildly up in the air. The reason this doesn't happen all the time with our toothbrush is because our internal models allow us to make estimates without the need for that feedback, because we've lifted that toothbrush—and objects like it—thousands of times. "A tradesman might have 10 different hammers for 10 different things," Ingram said. "He just picks it up and goes, because he's got a model of each hammer."

Motor models are highly egocentric, while sensory models seem to be more attuned to the outside world. "In general, when I move my arm, all these forces inside my joints are not really useful for me to be aware of," Wolpert says. "It's much more useful to be aware of external events happening—animals coming to eat you, predators, prey." The brain wants to get rid of anything extraneous. You do not want to be aware of every sensation going on inside your body. An efficient way to focus your attention on external events would be to subtract the more predictable components. "Anything that's unpredictable pops out more," Wolpert says. He used the example of the Gravitron at the county fair, the spinning room that produces centrifugal forces to flatten us against the wall. I remembered the ride well. The fun part was, after a minute or two, you could essentially move around without a problem. The forces were still limiting, and our bodies were still more or less confined to the perimeter. But you get used to it. Actually, your brain gets used to it. Those forces are effectively the same as an ice skater might make during a tight spin. But when you rotate your body, you don't feel them. "It's totally predictable," Wolpert said. "You don't want to be aware of things that are irrelevant."

Sensory feedback can come from only two sources: external or internal. What we see, taste, hear, smell or touch, or how our own body is moving. The brain does not know which is which. There isn't a label that says "This is an internal stimulus" and "This is an external stimulus." It needs to develop tricks for distinguishing the difference.

One of the ways is by reducing the feedback from our own movements in order to amplify the signals that arise from outside of us. The best way to understand this is by asking a question that Wolpert began investigating around the year 2000: Why can't we tickle ourselves?

He proceeded with the world's most sober investigation into tickling. Utilizing a robotic arm and a soft piece of foam—and perhaps a healthy appetite for sadism—he subjected participants to judge what felt more ticklish: when they controlled the robot arm and produced the sensations on themselves, or when they had no control over the arm. Not surprisingly, the subjects laughed a lot less when they were in control. And when the robotic tickle delivery was delayed (by 100, 200, and 300 milliseconds, thereby increasing the uncertainty), their sensation increased progressively as the delay increased. The results supported Wolpert's hypothesis that the brain is excellent at predicting its own movements, which has an attenuating influence on the corresponding sensation we feel. The reason, Wolpert believes, is so we are more attuned to external events. This can mistakenly lead to erroneous estimations. Wolpert and others discovered this when they asked people to hold out their index finger in front of a force transducer, which applied a small push. When the participants were then told to apply the same amount of force to a second transducer, they consistently pushed back harder and harder, in what could be described as an escalating "tit-for-tat" exchange. Self-generated force is perceived to be as much as *half* as strong as when it gets generated externally. It explains how shoving matches tend to spiral quickly out of control. Your sensation of being pushed is not even close to being aligned with your judgment for how hard you should push back.

"The brain is always trying to predict," Ingram said. After all, what is the first thing that babies do? "They don't walk or talk. The first thing babies do is reach and grab things. I don't think that's an accident." Humans are building internal models for their interactions with tools and the environment almost immediately. This is not the same for many other species. The first thing many animals do is walk. Their survival is predicated on evasiveness or pursuit. Ours is predicated on contrivance. "We have a lifetime of experience interacting with objects," he said. "It seems so easy, but it's actually really hard.

"Before modern medicine, women died in childbirth all the time," Ingram continued. "Forget how expensive brains are to run and to build. Having the female of your species dropping like flies, that's a pressure to make the brain as small as possible. And yet it is massive." Something took the brain from being relatively tiny to being the largest relative to body mass of any species on Earth, and as Ingram says, "We've got to assume that motor control is a part of that." The growth of the brain shadowed the increased human dependency on tools, he notes. "Look around now. Everything we can see is the result of the human ability to manipulate objects. Musicians and sportsmen, the incredible things they can do, is the result of having this big brain that is partly dedicated to motor control."

As Whitman wrote in "I Sing the Body Electric":

In this head the all-baffling brain,
In it and below it the makings of heroes.

It is one thing for Tom Brady's visual system to be able to remember where his receivers are located at all times. It is another for him to be able to make an accurate pass, which requires knowing how much force he needs to project the football to reach his target. The Behaviorists might have said he learned these throws by trial and error. Praise for the good passes reinforced the neural connections enough that the response occurs reliably, while punishment alternately weakened the connections. Edward Thorndike called this the "law of effect." "The function of intellect is to provide a means of modifying our reactions to the circumstances of life," he wrote in 1902, "so that we may secure pleasure." But by the middle of the century, thanks to Karl Lashley and others, Behaviorist hypotheses like that one began to wither from lack of reinforcement. Some felt their way of viewing experimental psychology was much too limiting. These were the Cognitivists. They did not think of the central nervous system like a telephone switchboard, transferring incoming sensory calls into motor output messages. There was more to it than that. One influential cognitive psychologist, Edward Tolman, instead likened the brain apparatus to a map, which lays out a schematic of the environment, with possible routes and paths, before a response is selected. He tested his hypothesis with various experiments involving rats in a maze.

In one maze, as depicted in the hook-shaped drawing on the following page, the rat entered at A and learned to find its food at G, after a slalom path.

The rat got to be quite adept at finding its food. Then after some time, the maze design was changed. A sunburst maze (page 286) presented the rat with a variety of counter paths. But the most familiar one—straight ahead, where C had once led to D to lead all the way to the food at G—was blocked.

Setting the rats free to explore, Tolman found the results

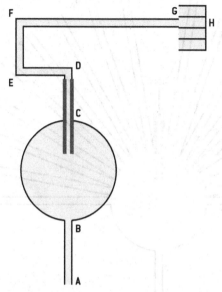

Tolman's classic Hook Maze, an experimental design in which a rat enters at A and learns to find food at G.

striking. The majority of them, approximately 34 percent, expressed a tendency to follow path number 6, a route that ran to a point only a few inches from where the food box had been in the original design. It was as if instinctively the rat knew the direction it needed to head, even though the setting had changed. The second-most frequented path was number 1, which ran perpendicular to the blocked straight-ahead route, just as the rats had learned to bang a sharp right from their experience in the original maze.

So what exactly had the rats learned? If the animals were conditioned to simply make responses, they would likely continually choose the alleys closest to the original straight-ahead path; those are the closest options to the one that had been reinforced. But it

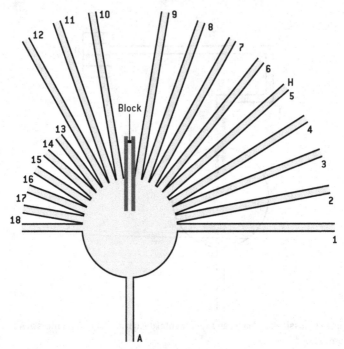

After training the rats on the Hook Maze, Tolman would change the setting to this sunburst pattern. Here, the rats again started at A, but their trained, straightaway pathway was blocked. Tolman discovered that the overwhelming majority of rats finally chose to follow path 6, a route that ended only a few inches from where the original food box had been. The second-most frequented path was path 1, a sharp right angle similar to the one from the Hook Maze. Tolman used these experiments to propose that the rats might employ a kind of "cognitive map."

appeared instead as though the rats had hazily learned *where* the goal was located. In Tolman's view, over time, they had encoded a topographic depiction of the space they were in. "Learning," he wrote, "consists not in stimulus-response connections but in the building up in the nervous system of sets which function like cognitive maps." His was a sophisticated and selective view of the central nervous system. He called himself a "field theorist," as with physics, someone who believed the brain was consolidating what it saw in its vision by assigning a vector along each point of an internal map. He was the first, but not the last, proponent of the inherent human GPS.

In the early 1970s, John O'Keefe, then a researcher at McGill University, decided, almost by accident, to place microelectrodes into the CA1 region of the hippocampus of a rat. It was the memory-encoding brain region made famous by another McGill luminary, Brenda Milner, and her work with Patient H.M. But O'Keefe was not as much interested in episodic memory. He simply wanted to see how the hippocampal cells fired. He found, studying rats in a cage, that the neurons fired differently depending upon where the animals were located. Some cells fired when the rat was in the southwestern portion of the enclosure, while others fired when it was in the northeast. For O'Keefe and his student Jonathan Dostrovsky, it became possible to effectively tell where the rat was standing in its cage just by looking at its pattern of neural activation. They thought of these neurons perhaps as "place cells," responsible for dictating to the rest of the brain where the body was placed. Large groupings of place cells were found to be responsible for different regions of space, producing an entire place field.

Without portions of the hippocampus, animals have been shown to be unable to perform a spatial task. In other maze designs, including Richard Morris' water maze, a rat might be placed in the center of a ring with a number of alleyways sprouting outward, each containing some bit of food at the end. The rat has to learn to get the food down one alley, return to the center, and then choose another alley. Over time, it will get to be quite skillful at remembering which alleys it has already visited and which ones it has yet to target. And rats with damage to their hippocampus cannot do this as well. They often will recall the task, run down one alley and promptly forget which alleys they should go down next.

Along with place cells (found in human studies of epilepsy patients as well), the hippocampus and the nearby entorhinal cortex also appear to house certain head direction cells, which fire whenever an animal is looking in a particular direction; boundary cells, which signal distances from walls in the environment; and grid cells, which seem to fire in patterns that tell the animal the distances it is heading in any given direction, independent of context. The latter discovery, made by Edvard Moser and his wife, May-Britt, in 2005, was confirmed in humans in 2013. It completed the picture of the brain's internal navigation system, that intuitive sense of where we are and how to get where we are going. In 2014, the Mosers and O'Keefe shared the Nobel Prize in Physiology or Medicine.

With place cells in the hippocampus, and grid cells in the entorhinal cortex, it has been suggested that a third brain region, the posterior parietal cortex, functions as a sort of navigational translator for the motor system. Indeed, the parietal region's influence in our ability to integrate sensory information and form an understanding of extrapersonal space is perhaps best understood by the impact felt when it is damaged. In the late 1970s, an Italian neurologist, Edoardo Bisiach, observed a group of Milanese patients, all with injuries to their right parietal lobe. When he asked them to imagine they were standing in the middle of the Piazza del Duomo, facing the world-famous cathedral, he found that they could all clearly remember the buildings on the right of the square but not the left. When he asked them to switch perspectives, imagining themselves looking out from the duomo toward the square, they could accurately identify the buildings they had previously forgotten, while blanking on those they had just recalled.

Without a functioning hippocampus or a posterior parietal cortex, we might be, in effect, bodies lost in space. Even Sherrington, with his focus on the spinal cord and reflex action, recognized the necessity of the cortex for the most critical aspects of behavior. Throughout all his experimentation on animals, cutting off the connection between the brain and the spinal cord, he had also cut off the connection between that animal and the rest of the world. "By a high spinal transection," he said, "the splendid motor machinery of the vertebrate is practically as a whole and at one stroke severed from all the universe except its own microcosm and an environmental film some millimeters thick and immediately next its body."

Given the effect of a damaged parietal lobe on memory and orientation, is it unreasonable to suggest that a skilled athlete's knowledge of the field might be related? If you, like many, have wondered how exactly a wide receiver knew to plant his feet down at the precise point before tumbling out of bounds to make a catch, while keeping his eyes firmly trained on the ball, then maybe you have thought about the influence of the brain's inherent GPS, too.

It would seem to fit a report from Japan, where a former national soccer star, Hidetoshi Nakata, who was known for possessing a legendary "killer pass," revealed some years ago that he scored a 19 out of 19 in the Block Design subtest of the Wechsler Adult Intelligence Scale-Revised, a portion of the popular IQ test that is said to challenge spatial ability.

In 2008, a team of Japanese researchers followed up by testing

Block Design on 31 male rugby players. They found that, on average, they scored significantly higher than a standardized sample of 163 Japanese men on the same test. The authors took it as a sign that visuospatial reckoning is a distinguishing characteristic of top-tier rugby players. Two years later, another Japanese research team put 20 players from the same rugby club inside an fMRI. They gave them a mental rotation task, which has been shown to correlate with the Block Design subtest. The subjects were shown 30 pairs of 3-D drawings and had to judge whether the pairs were mirror images or the same by pressing a button. Though neither the experts nor the novices proved to be very good at the task, their strategies for doing it—as shown by their brain activity—appeared to differ. The rugby players, for example, displayed considerably more engagement coming from the right superior parietal lobe as well as a portion of the visual cortex. The researchers hypothesized that those brain areas can function in spatial transformation of a scene to a third-person perspective—the kind of viewpoint that, in a fast-moving sport like rugby, might allow players to better keep up with the widespread motion. The players, they wrote, could be seeing the field as if from a bird's-eye view.

O'Keefe has continued to study the characteristics of place cells. He has found that visual information is not even required for place cells to fire, as long as the animal was still moving in the dark. The body's simple exertion informs the hippocampus where it is. Something similar has been noticed in humans. In a laboratory, researchers asked how we are able to find our way in the absence of light. The key, they found, was having not only a visual memory of the path we are traveling but also our own interoceptive representations of that path—information gleaned just by the motion of our actions and the signals generated by our muscle spindles—which supports the idea that the brain is able to form a picture of our surroundings without the need to see them.

Recently, a walk-on lineman at the University of Southern California named Jake Olson made headlines for delivering the final snap on an extra point in a 49–31 win. It was one play in a blowout victory, but he was mobbed on the field. Olson has been completely blind since age 12.

After the game, as usual, he chalked his skill up to the familiar procedural concept that often seems to hog the motor spotlight: "muscle memory." In this case, he might be right. Long-snapping might not be precisely analogous to the dynamic, unpredictable requirements of a quarterback. But there is enough evidence to suggest that, for people like Tom Brady, "seeing" the field isn't everything.

9.

A PARALYZED MAN WHO MOVED

THE FUTURE OF MOVEMENT

C lose your hand. Bring your arm down. Bring your arm in. Take a bite."

The technician on the lower level of the Louis Stokes Cleveland VA Medical Center enunciated his directions clearly and slowly. A pretzel stick wobbled between the pallid thumb and forefinger of Bill Kochevar, whose paralysis extends from the neck down. He could see the stick in his grip but could not feel it. Although in a second, he would taste it. He alone would move it to his lips.

By virtue of the sublimity of their motor systems, and the dilution of their N, many performers can make the impossible seem easy. Unfortunately for many others, the opposite is true. But that gap along the grand continuum of human movement is slowly shrinking. New hope exists for those with even the most devastating motoric impairments, and much of it happens to reside in the revived right arm of a 57-year-old tetraplegic. Now Kochevar can grip a pretzel or a sponge or a spoon and manipulate it with his intentions. A limp limb suddenly springs back to life. His thoughts, once restrained, can again be transformed into bodily action.

Kochevar is tall, with green eyes, a barbed wit and a fierce competitive spirit. The latter can conflict with his lifelong fandom of the Cleveland Browns. In 2006, while cycling in a race to benefit people with multiple sclerosis, he slammed into the back of a stopped mail truck and fractured the C4 vertebra of his spine. It had been raining that day and visibility was poor. He was left with some sensation at the top of his right shoulder, very limited command of a few fingers on his left hand and mostly free range of his neck, but nothing else. His right arm, as well as his entire torso on down to the toes, is locked in paralysis. He guides himself in a power wheelchair using his tongue. Everything else requires assistance. But in 2013, researchers from Case Western Reserve University and Cleveland's VA hospital approached him looking for a participant for a pioneering experiment. Kochevar, the former navy radioman now confined with little to lose, seemed like an ideal candidate.

The process was not going to be easy. Kochevar needed to undergo surgery to place a small metallic electrode patch into a portion of his motor cortex, and then, in a separate procedure, get even smaller stimulation electrodes implanted into the muscles of his right forearm, elbow and shoulder. These were inserted with long hypodermic needles. The limb electrode system, called "functional electrical stimulation" (FES), pings a pattern of currents into the muscular nerve to produce a coordinated movement. Until recently, the pattern was commanded externally, by either a computer or a pacemaker-like device that was implanted near the chest.

But the goal was to get FES to work in conjunction with the cortical patch, also known as a "brain-machine interface" (BMI). With the two together, the scientists felt they might have found a sort of work-around for severe spinal-cord injuries. With cervical damage, for example, the signals from the cortex could not travel down the spinal cord to stimulate the motor neurons in the periphery (neither could the afferent sensory or proprioceptive signals travel up). The motor cortex on its own was not impaired; it was just sending signals to a dead end. The electrode patch would serve as a recording device, eavesdropping on the signals and diverting them, along an external cable, to a neural signal processor, a computer, to decode the messages.

These messages would then be fed to the FES, which was designed to interpret the signals and confuse the muscle nerves into thinking the electrical stimulation it is providing is not artificial but rather naturally flowing from the brain. The nerves might then jolt to attention. The arm would be mobile again. When the research team, led by Bob Kirsch and Bolu Ajiboye, brought this idea up with Kochevar, who then lived at home with his father in a Cleveland suburb, he liked the sound of it. He said it seemed "Star Trek-y." But he hesitated to immediately agree to it. Kirsch expected Kochevar to decline for the same reason he had heard from a lot of prospective patients: that he did not want to be first.

But being first was actually what intrigued Kochevar the most about the idea. "I'll be the first in the world to connect the two systems together," he told me. "That sounds pretty cool—to be first." His dad had a few more reservations. "He said, 'Do you really want to do this?'" Kochevar recalled. "I said yes. He said, 'Why?' I said, 'Because it's research.' Somebody has to do research. If nobody does it, things don't get done." Kochevar called Kirsch back and agreed to proceed.

"Were you scared to be first?" I asked.

"No, because other people before me had done the BrainGate—put the things in the head," he said. "And other people had done the FES. I was just the first to connect them."

The experiment, which began with the cortical patch surgery in December 2014 and concluded with the second of two FES surgeries in September 2015, has been considered a phenomenal success. When details of the work were first published in the journal *Lancet* in March 2017, it was quickly hailed as "groundbreaking." *Time* magazine labeled it a "major advance." A video produced by Case Western showed Kochevar drinking from a cup of coffee and fork-feeding himself mashed potatoes. It was shown on CNN, CBS News and the BBC.

When I visited Kochevar one spring afternoon, he had been working with the system for almost two years. He still arrives at the lab at least twice a week for three-hour research sessions, and I joined him as he worked. He rolled into a lab room with violet walls and laminate flooring just after 1:00 P.M., in a green Henley T-shirt and gray gym pants. He looks forward to his research days, he said, and we chatted as two of the technical support staff members got everything set up, a process that can take half an hour. He spoke slowly, with long pauses in between sentences, taking his time. He has plenty of time. "I've had fun doing it, and it's awesome that things work as well as they do," Kochevar said. "Every time I do something new or different, it amazes me that I can do that type of stuff."

I asked him to describe what it was like seeing his arm move again for the first time.

"I really thought initially I was never going to move again," Kochevar said. "I always hoped that I would regain some movement, but I didn't. I came to the realization it was a complete injury and I was going to be like this for the rest of my life. But BrainGate changed my thinking." BrainGate is the name of the consortium involving Kirsch and Ajiboye and more from Case Western, along with others from Stanford, Brown, Massachusetts General Hospital, and Providence VA Medical Center, dedicated to using brain-machine interfaces to resolve motor impairments.

The FES requires that the bundles of hair-thin wires protruding from three places along his arm are responsive and connected, via cable, to a Universal External Control Unit, a gray box no bigger than a wireless router. Kochevar's right arm is placed on a scissor-lift stand that can robotically assist his shoulder, a body part that, perhaps owing to the muscle atrophy after years of disuse, seems to be the most stubborn toward the intervention effort. The stand is also solely controlled by his brain signals. It moves only when and how Kochevar intends it.

But the brain interface is more complicated, and I watched as the technician, Brian Murphy, carefully replaced the antiseptic Bio-Patch disks, darkened from blood and sweat, around the two thick bolts protruding from the top of Kochevar's skull. Kochevar says he does not feel them. But his hair has yet to fill completely around the holes where a neurosurgeon extracted parts of his skull to attach the patch. From the bolts, heavy cables snake down the back of Kochevar's wheelchair, where they plug into mounted amplifiers, to intensify the recorded neural firings and digitally convert them to be read by the computer. Those feed into two neural signal processors on a trolley that contains three monitors and Murphy's laptop, along with a dizzying amount of red and green wiring, all responsible for decoding the subtle bursts of activity from neurons once in command of Kochevar's now idle limbs.

When I saw the electrode patch (a spare one), I was stunned. It was no bigger than a red pepper flake. It was four millimeters by four millimeters in area. Two of them could fit quite comfortably on an aspirin. On the patch were 96 even smaller dots, each one of them a separate recording electrode, and each of those pings the brain approximately 30,000 times per second. The patch can record from a population of several dozen to more than 100 neurons. It is spread across enough real estate to record and restore the signaling of some abstract motions, like reaching and grasping, but not individual finger movements. Kochevar could not sit at a piano keyboard and start playing. But the motor function largely returned to his limb as if reawakened from hibernation. With it, he has been granted two grasping techniques: a collapsing of all four fingers into the palm, and a pinching between the thumb and index finger, as you would push the button on a joystick. But those two maneuvers, thanks to the synergistic nature of our movements, can account for about 90 percent of our hand function. The only thing Kochevar needed to do is imagine his right arm moving and, as if by magic, it responded to his thoughts again for the first time in nearly a decade.

"You ready, Bill?"

Murphy had the setup completed and was awaiting Kochevar to take another test-drive of his now-animated imagination on his own. The first task was simple. A voice would call out which direction to move the arm, and Kochevar would follow. His wrist was braced and his arm rested on the sling atop the scissor lift.

"Yes, I am."

"OK, here we go."

The first vocal command was "down." Kochevar stared at his arm. Suddenly, slowly, it began to lurch downward. After about five seconds, the next command: "up." His brow furrowed. He stared at the arm, trying to raise it as Luke Skywalker might employ the Force. It eventually rose to a point about eye level and remained there, wavering slightly, awaiting the next command.

After about 20 trials, the voice stopped. Kochevar had had some difficulty with commanding the arm to retreat back down, but otherwise seemed to get the hang of the instructions to move up, in and out, and open and close his hand. Before the accident, Kochevar worked in IT for an insurance company. Today it is a computer that functions as the middleman between his thoughts and his actions. While setting up the session, Murphy monitored the channels of information coming from the electrodes, which appeared on one of the screens as 96 individual graphs of various widths and densities. The denser graphs represented the firing patterns collected from the electrodes that were closer to the source of the neural command he was giving. Another screen displayed data collected by motion sensors that are attached to the scissor lift, conveying where Kochevar's arm is in space. Because Kochevar has neither sensory input nor any proprioceptive information telling the brain where the arm might be, it is effectively making motions in the dark. Like Ian Waterman, he relies on his vision to inform him where his limbs can be moved.

And to move, all Kochevar needs to do is think. His eyes can be open, and most of the time they were. He looked around the room, maintaining a stoic expression somewhere between task fatigue and complete disinterest. "I don't have to think hard," he said. I asked him if my presence standing next to him posed any distraction to his thoughts, and he chuckled no. Certain movements looked natural, such as when he opened his palm while making an upward point, as if casually hailing a cab. Others, like his ability to move his arm downward, took several trials before he was able to master it. For one of the tasks, Bill Memberg, a technical support staff member, placed a pencil with a large circular sponge at the end in Kochevar's hand.

"Close your hand," he said.

After a moment or two, Kochevar's fingers gently wrapped around the thin pencil shaft. He did not need instruction for what to do next. In slow, stilted movements, he bent his elbow at the joint, raised his shoulder up, then down, positioning the sponge just above his glasses. He craned his neck down and nuzzled his forehead onto the sponge's bristly side, like he was relieving an eight-year itch. A smile creased his face.

"Ahhhhhhhhhhh."

The reason some movements return more sluggishly than others is because the targeted neural population in his cortex is constantly shifting. Though the electrode patch is fixed in one place—a place known to be among those responsible for arm control—the scores of neurons they are recording from can vary. This complicates the algorithm for decoding. The researchers did not fully appreciate the dynamism of the cortex until they tried recording from it. "The neurons move around," Ajiboye said. "The population of neurons from day to day is somewhat different. Some may be the same, and some may not." It was conceivable, when Kochevar was struggling to move his arm early in the session, that the algorithm for a downward motion had shifted from his previous session. The neurons were different. With hundreds of millions of neurons firing signals down a functioning spinal cord, the variability of a few dozen of them on a given day is negligible. But in such a highly specified range being picked up by the recorders, precision was crucial. Murphy and Memberg needed to ensure that the signals they were picking up were the right ones to translate into the desired action.

There were other surprises. When Kochevar first got the BMI implant, before he received the FES system for his arm, he used his thoughts to control a cursor on a monitor. To try to establish some of the algorithms early, as a starting point, Kirsch and Ajiboye created a "virtual arm" that Kochevar could control on-screen. He even wore 3-D glasses to complete the effect. "It seemed 3-D to him," Ajiboye said. "And we tried to figure out what the mapping was. So, like when he wants to move his arm, his brain cells fire like this, and when he wants to move his hand, his cells fire like that." When they finally got the FES up and running, though, they tried to use the same algorithms to control Kochevar's arm as he reached and grasped objects in reality. But it didn't work. His arm remained inanimate—no matter how hard Kochevar thought—until the researchers remapped. "The perception of the end-effector, whether it's a virtual arm, robotic arm or his own arm, makes a difference as to how his cells will process the movement," Ajiboye said. In other words, context may be key. A virtual environment was not the same as the real thing no matter how lifelike the simulation might have seemed. "We're trying to understand the context of movement," he said. "Where he's looking, what is in his environment, all those contextual cues may have some effect."

I asked Kochevar what he thought about when he tried to make a movement. Was his arm floating in space? Was he gripping a football like a member of his beloved Browns? He said no. "I just think about the movement," Kochevar said. "If I just think a little bit about it, it just moves." Initially, he felt like he really had to concentrate to imagine the movements he wanted to produce. But lately, the thinking has gotten easier. "It pretty much just comes on its own," Kochevar said. "I'm still probably thinking about it—the brain is still sending the signals to do that kind of stuff. But I don't really notice I'm thinking a lot about it." This could be chalked up to practice, to the brain's plasticity re-forming the channels it once relied upon for skill or to Kochevar gaining more familiarity with the tasks and the system. Ajiboye said it is equally likely that they are simply improving at calibrating the signals the brain is generating. "It's unclear if he's getting better or our decoding is getting better," he said. "We're both improving."

"My brain signals change from day to day," Kochevar said. "It's hard to connect the two up and make them work perfectly. We're getting better at it, but it's always an ongoing process."

The next task was Kochevar's favorite, because it involved food. Memberg held a pretzel stick out in space and provided verbal instructions for Kochevar to follow in order to reach out and grab it. "Close your hand," Memberg said, placing the stick in between his opened fingers. The fingers slowly clenched around it. "Bring your arm down. Now in." The stick made its way toward Kochevar's face. He ate it in four chomps, deftly avoiding biting down on his fingers, which, of course, he still could not feel. But a proud look washed across his face: It had been a good day. I asked what he hoped the experiment could eventually allow him to do. "Just all sorts of daily tasks," he said. "Stuff that I can do in my daily life that I have to have help with right now."

"I would like people to be able to be reasonably independent from caregiving," Kirsch told me. "It doesn't have to be standing on one hand while playing the piano. It's not going to be like that, not for the foreseeable future. But if we can get them to the point where they can be semi-independent, they can go for a few hours at a time without somebody doing something for them. Ultimately you'd like to do more and more, but that is what I would say is my nearer-term goal."

Bob Kirsch has been working at the Cleveland FES Center since 1993, and the funding for an experiment involving BMI and FES in combination began as far back as 2005. A private company called Cyberkinetics raised $40 million to push BrainGate and its concept for clinically viable brain patches into the mainstream. But by 2008, it had exhausted all its money. Many thought the idea was destined to languish as a pipe dream. It is easy to look back now and say that the technology was about a decade too early. There was a shortage of willing patients. There was also a shortage of willing surgeons. Because the brain implant is percutaneous, it is susceptible to infection. The ideal FES system was another fully implanted pacemaker-like device that was placed near the collarbone, with wiring leading down the muscles from within the arm. But any prosthetic implant—hip, knee, heart, artificial limb stimulator—is frequently walled off from the body by a Saran Wrap–like material called "biofilm," which can be prone to collecting bacteria and can quickly lead to infection. People with new hips, for such a reason, are encouraged to take antibiotics before visiting the dentist. You can only imagine how a doctor might feel about a patient with two raised pedestals sticking out of the top of his head.

When the National Institutes of Health swooped in to salvage BrainGate in 2008, more impediments arose. Even after finding a willing patient, Kochevar (who would not receive a fully implanted FES system in his chest but rather would receive smaller wire implants into his arm muscles), there was the question of what they might find after opening his skull and attaching the patch. Before the procedure, Kochevar was sent into an fMRI to measure the activations in his motor cortex as he imagined the various movements he was hoping to soon make on his own again. The imaging recordings gave the researchers an idea of where to place the patch. But, Kirsch said, that is hardly a foolproof way to distinguish 60 to 100 of the critical neurons upon which they hoped to eavesdrop. "I'm just going to say it's really hard to get intention from somebody who doesn't actually move," Kirsch said. "When we recorded those signals, was he really, really thinking about moving his arm? Or was he daydreaming about Marilyn Monroe? Was he really even thinking about it? It's just tricky."

They had an idea where to place the patch based on Penfield's homunculus. "But it's very fuzzy," Kirsch said. The representation of the arm within the motor cortex is widely distributed, and their patch was narrowly confined. They were landing on an aircraft carrier in the middle of a dark ocean. And they were basing their aim on the testimony of a motionless man inside a scanner bore. "We have both arms [tied] behind our back when we do these things," he said.

Kirsch had another big concern: Could the motor cortex even produce the right commands to move again?

It had been nearly a decade since Kochevar's accident. His arms, his legs, his entire body, had been isolated from his brain for so many years it was an open question whether the brain still remembered what it was like to be in control of it. It was an open question whether the body would know what to do with new top-down commands as well. Kirsch understood that plasticity, and the motor cortex, in particular, gets continually reshaped by experience. When someone develops a focal lesion, for example, that area can be functionally reorganized so that another area of the cortex handles the motoric responsibilities of the affected region. So, Kirsch wondered, after so many years, what portion of the motor cortex had been repurposed for something else? To what degree could it possibly have retained its functional eloquence or dexterous precision without any use? Had any of Kochevar's hard-earned abilities—as a cyclist, a trained technician, an amateur French horn player—remained frozen in time, waiting for the outlet through which it could be released? And when they did record from the cortex, what might they hear?

The first ever to listen was Herbert Jasper, a pioneering psychologist in the work of EEG. In 1953, Jasper and a collaborator, David Hubel, designed intracellular microelectrodes out of tungsten that could be implanted in animals to overhear the electrical activity from single neurons as they performed tasks. They could then uncover patterns. A single neuron is a hard thing to fathom. One hundred of them would fit on the period at the end of this sentence. Their connections are forged by the action potentials—the firing of a signal down the axon and out to the dendritic tendrils of an adjacent neuron. Neurons either fire a signal or they don't; there is no ambiguity. Just as you cannot half-fire a gun, it is all or nothing once they pull the trigger. But how that activity correlated with behavior or cognition used to be unclear. Not long after Jasper's innovation, Edward Evarts at the National Institute of Mental Health refined the technique enough to fundamentally shift our understanding of motor control. By conditioning monkeys to release a telegraph key after a cue, he could detect differences in the rates of when neurons *decided* to fire. Some neurons increased their rate with the force generated in the movement, while others increased in order to hold still. The activity typically began as much as 50 to 150 milliseconds *before* the onset of the movement. And single neurons in the motor cortex were shown to correspond with specific parts of the body, such as the hand or wrist. When Evarts published some of his initial findings in 1967, he credited Charles Sherrington. "Sherrington had written that the problem of whether the discharges of motor cortex neurons represents a step toward psychical integration or, on the other hand, expresses the motor result of psychical integration or are participant in both is a question of the highest interest, but one which does not seem as yet to admit of satisfactory answer," Evarts wrote. Decades later, he had successfully pointed researchers toward the way that neurons generated signals to produce specific guidance for the movements of specific body parts before they can be executed.*

Neurons could vary the rate or intensity of their signaling, and

---

* Continued research on the activity patterns of cells has produced evidence that even recognition of a certain face can prompt a neuron to fire. In 2005, a single neuron in an epilepsy patient was shown to respond only to images of the actress Jennifer Aniston but not to images of the actor Brad Pitt. More recently, researchers have pinpointed about 10,000 neurons whose firing patterns are implicated in the recognition of human faces. Interestingly, the cells were found in the medial temporal lobe, not the fusiform gyrus.

that could be coded as the instruction for behavior. But a debate intensified about the nature of this control of movement. Did the neurons in the motor cortices simply provide the kinetic information about planning and force, and leave the rest of the kinematics up to someplace else? Or could neural firing patterns be responsible for more information? Some of this would not get elucidated until 1982, when a young researcher at Johns Hopkins named Apostolos Georgopoulos and his colleagues recorded from the motor cortex of a monkey as it moved a joystick toward eight targets arrayed in a circle around it. Different firing patterns, they found, correlated to different directions of movement, as if each neuron was "tuned" to a specific direction. Single neurons could fire for multiple directions, but activity was strongest for a "preferred" direction and weakest for a movement the opposite way. There was a complete gradient of firing intensities around the entire circle, representing all directions. Cells with similarly tuned direction preferences were found across several different sites in the motor cortex responsible for arm movement, and adjacent cells could also have completely different preferences. Georgopoulos took this to suggest that the activity pattern of an *entire population* of neurons was what provided the signal for movement. His findings became known as the "population vector hypothesis," summarized by the idea that motor output is effectively the result of the summed activity of a large population of neurons in specific regions of the cortex.

Dagmar Sternad once told me that peering too closely at the neurons in the brain is like trying to understand clouds by their individual water molecules. "It is by the way that they hold together that we understand the patterns forming," she said. Indeed, Georgopoulos' experimental findings at once seemed to confirm and oppose the earlier discoveries by Evarts. For years, the all-or-nothing characteristic of neurons fed the notion that they were distinguished by their location in the brain, as if by neural zip code. But, as it turned out, they may have different dialects and homespun quirks, too. Individually, they were like snowflakes, composed of and responsible for a random combination of . . . something. But motor behavior is reflected in the connectivity patterns of the neurons in groups. If their dialects could be interpreted, and enough neurons were gathered, it was possible to predict the direction of a movement just by eavesdropping on what they were saying.

Three decades later, that is how Bill Kochevar is able to move again. Lee Miller, who runs a lab at Northwestern that also develops neural interfaces for patients with spinal-cord injuries, told me that Georgopoulos' fundamental ideas are what research teams around the world are relying on to decode and bring movement back to either paralyzed or prosthetic limbs. "He proposed that if it were possible you were able to record lots of neurons at the same time, you could predict where the arm is going and actually use it as a control signal," Miller said. "That's exactly what just about all BMIs are doing these days—whether it's movement of a cursor on a screen or the endpoint of a robot. It's all related to the idea that you can predict a direction of motion by looking at these neurons."

Researchers have used that idea to get patients to maneuver cursors on a screen, operate computer programs, compose messages or control a robotic prosthetic by just thinking about the corresponding movements required. Those intentions activate the neurons in the motor cortex just as any normal movement would. There were people—as opposed to muscles—now doing the listening. "People have spent the last 20 to 30 years building decoders—algorithms that would match the brain activity to some parameter of movement, whether that is direction, velocity, joint, intention," Ajiboye told me. "We're basically applying that neuroscience knowledge to human participants."

They were just hoping that their aim was pure. The goal was to tap into the natural circuitry of neurons that relate to hand movement and let Kochevar extort that circuitry seamlessly and effortlessly. "What he's told us is he is just naturally thinking about his arm movement from Point A to Point B," Ajiboye said.

"And what does that tell you?" I asked.

"It suggests that the signals we're recording have a natural correlation to actual movement," he said. "We're harnessing the circuits that were related to movement prior to the injuries. These have been preserved. And our electrodes are in the right place."

Kirsch added, "The cortex is pretty adaptable, but there is also a hard copy in there. Bill has been paralyzed a long time, and you might think maybe that part of the brain is going to be used for something else. But it still remembers how to control the arm."

Game 3 of the NBA Finals was about to start later that week in Cleveland, not far from the medical center. I could not resist but draw a hypothetical parallel between what Kirsch and Ajiboye had achieved with Kochevar, what they had learned already from Kochevar's motor cortex, and what LeBron James and Stephen Curry were about to do with theirs on the floor of Quicken Loans Arena. I could not help but think that Kochevar's strained grasps today might eventually evolve into a complex skill tomorrow, as researchers inch along the continuum from expertise to impairment, and back again. I asked Kirsch if Kochevar's progress gave him the same aspirational feelings about the future.

"Well," he said. "He's moving."

One week later, I decided to take a drive west out of Manhattan along Interstate 80 toward a place called Flanders, New Jersey. It is not a very populated area of the state; land is protected as part of the Highlands, a watershed ranging from western Connecticut down to eastern Pennsylvania that supplies nearly six million people with drinking water. Because of that, it is largely a farming region. In the 2016 presidential election, it was a Donald Trump region (51.2 percent). I visited because it is also a baseball region.

Jason told me he had recently sold the uHIT kit to a youth baseball academy there called In the Zone, which caught my attention. After two years of hassling with Major League teams about their usage, it had become clear that deCervo was looking to pivot in another direction. The clubs had deep pockets, but turning the organizational battleship, as Jason liked to say, was proving even harder and more incremental than they expected. They had introduced the potential of neuroscience to Major League Baseball. They felt they had done what they could for the motor cortices of the professionals. But they also could not wait around forever for teams to show anything more than a lukewarm understanding. They had to be businessmen now.

"The business model was originally, we said, all right, Major League teams have a lot of money. They're clearly going to want cognitive stats that can make their players better, and we can measure this stuff. So we'll sell this as a service." This was Jason a few weeks earlier. We had sat in a glass-walled conference room on Sixth Avenue with a guy named Anthony, a fast-talking venture capitalist in gray jeans and sneakers who seemed to be angling to get a piece. He had recently bought an independent league baseball team in Connecticut and was planning to use the team and the stadium as a technology trade showcase: apps, games, sims, VR, AR, whatever might stick. But neural was new. Neural was different.

"You guys are right in my sweet spot," Anthony said.

The sweet spot he was referring to was not necessarily professional baseball. No, Anthony wanted to help take deCervo to the masses. And the more Jason was hearing, the farther he leaned back in his chair, the more he liked. He had come into the meeting to pitch Anthony on advertising within the simulation, because Anthony had cofounded a successful digital ad firm. There were banners on the outfield walls that could be sold a different branding than "deCervo." But Anthony quickly dismissed the advertising angle as an insignificant fraction of deCervo's ultimate potential.

"That's easy revenue," he said. "But what's going on with kids today that are between 8 and 18 years old is they grew up in a generation of needing to be seen and recognized by their social following. These kids, right or wrong, have been told that they are the best baseball players in the world. They're uploading all these videos their parents took of them up on social sites. I think what you guys can create here, I think there's an enormous market for that same 12-year-old or 18-year-old to showcase that they are the best at pitch selection and accuracy and reaction time."

Scouts would be using deCervo's app on their phones checking to see how kids scored in the game, Anthony said. "A scout can see, 'something is going on here.'"

"It's not the million-dollar arm," Jason said. "It's the million-dollar brain."

"One hundred percent. You've got a lot of legs. You've got revenue legs." Anthony paused. "And your background is . . . ?"

"Physics, aerospace engineering and neuroscience engineering."

"There ya go."

"And music," Jason said.

"Music?" Anthony said. "Neuroscience and music?"

"Actually, they mix well together," Jason said. "That's how this all started. I was at Columbia, I'd switched from aerospace into neuroscience engineering, and I didn't know anything about neuroscience. I started thinking, though, that a nice experiment might be looking at musicians' brains. What's different about them? This all started by finding that there was something about the neural circuitry of musicians that was different than other people."

It was true, in a way. It had started with the brain and music, the synchrony of the nerves working, as Galen once wrote, like "conductors, bringing to the muscles the forces they tap from the brain as from a spring." The hands, those same hands gripping a bat so tightly the blood recedes out of the fingertips, those hands "are an instrument, as the lyre is the instrument of the musician, and tongs of the smith." Eighteen hundred years later, a person could still be stunned to hear that a simple neuroimaging contrivance could relay the inner workings of that same conducting mechanism, with fine enough detail to be able to distinguish differences in its coordinated output, like an X-ray of the mind.

"Can you predict one person who is going to have a better reaction time than another?" Anthony asked.

"Yes," Jason said. "I mean, the whole bunch of studies that Jordan and I did together were looking, starting with EEG first, can we measure the signal with EEG, then how does that relate to physical reaction time, then the decision signal—when they decide not to do something—do those decision times differ between baseball players and non-players, and how do they differ? Then unpacking what timing aspects are different."

"Have you done that?"

Yes. Jason could have brought up the fusiform gyrus, the region that picks up baseballs like bird-watchers spot a warbler in the bush, or the supplementary motor area, the timing region that lets great hitters hold back their swing. He could have mentioned the role of the AON, the neural belt that changes as we move and reflects those changes into how we move and how we learn. He could have mentioned the flawed study on Babe Ruth, which stood for decades as a paean to the Bambino as a physical marvel while perhaps ignoring his apparent amblyopia. He could have mentioned neuromotor noise, the enigmatic N, afflicting all our movements, although not everyone's in the same way.

It was not the setting to get ensnared in the scientific weeds. Jason Sherwin and Jordan Muraskin had all but left that part of their careers behind them. But I did wonder if the next generation was going to start thinking about hitting not in terms of miles per hour, but in terms of milliseconds. If decision time was going to become a new item on the back of the baseball card. If the brain's role in hitting was going to finally get its due beyond a Yogi-ism.

When I called In the Zone, the owner, Marcus Ippolito, told me to swing by. He never said explicitly whether the system was up and running, or if he was even using it, and I never explicitly asked. I figured I would take a drive and find out.

I pulled up to 37 Ironia Road, a 32,000-square-foot warehouse with several loading docks. In the Zone leased the space in the back. I didn't realize it at the time, but I would find out later that there is another youth baseball academy, Complete Player School, at 27 Ironia Road, two buildings down. It was becoming obvious where people spent their summer weekends here. Ippolito was not there yet when I arrived, so I sat in the waiting area. A few parents were waiting on their kids to finish their lessons. I understood then what Anthony was saying about the youth baseball machine: travel teams, hitting academies, personal trainers, scouting apps. There seemed nothing out of the realm of plausibility that somebody might purchase to improve a child's chance of getting noticed, make the differences stand out. "I spent $250 for an aluminum bat," Judith Sherwin said to me once. "Why not spend on something that could teach him how to hit better?"

I overheard the parents plotting their trips to upcoming softball tournaments.

"Sunday's tough, 8:00 A.M. You staying over Saturday?"

"We're only about an hour away, so it's not really worth it."

"Same here. But next weekend we're going to stay over."

"Yeah? Where you staying?"

"I found a place called the Inn at Voorhees. I booked that one. Didn't you say you were there?"

"Just the Saturday night. I really wanted the Fairfield Inn, but they didn't have any rooms. I've got so many points. There's also a SpringHill Suites, I saw."

"Voorhees is a ways."

"If it's 8:00 A.M. that's not crazy."

"We'd have to be there at seven."

Just then Ippolito came in. We went into his office and he explained his vision for how deCervo might fit into his plans for a new facility replete with a "hitting lab" and a video screen at the end of each batting tunnel. "At the end of a lesson," he said, "we'll have the kid go over there and maybe finish up by doing some ball recognition work for a few minutes." Plus, the kids seemed to like it.

Ippolito brought me into the training area, where six retractable batting tunnels were stacked up all in a row along an Astroturf floor. The Gumby-colored walls and industrial pendant lights gave the space the dim, shadowy impression of an auto repair shop. Next to one of the cages on a stand was a flat-screen computer monitor and the familiar uHIT Virtual hardware package. Three teenage boys were milling about nearby. Ippolito asked if they wanted to play. He powered it up and asked them which pitcher from the eight preprogrammed into the system they wanted to face.

"Kershaw," they said, almost in unison.

As each took his turn against 77-miles-per-hour fastballs and parabolic curveballs, I heard them teasing one another about incorrect Go's and incorrect No-Go's and accuracy percentages. I chatted with Ippolito for a few minutes, but when I turned back, the kids had moved on. They had moved on to bunting practice. I hung around to watch another boy take batting practice in one of the far cages. The instructor was soft-tossing him underhanded, but the ball exploded off his bat. *Pop pop pop.*

ACKNOWLEDGMENTS

This book is dedicated to the memory of Bill Kochevar (1961–2017), who passed away while the first edition of this book was in the final stages of production. Thank you for allowing me to spend time with you, ask questions, and learn about your courageous fight to improve the lives of the motor impaired. Rest in peace.

Lydia Shamah is not one to frequent the sports section. But, thanks to some incredible stroke of serendipity, she came upon my article about Jason Sherwin and Jordan Muraskin that ran for SB Nation and recognized its book potential well before I did. I have to believe she would make a good baseball scout, if she knew anything about baseball. I owe so much to Lydia and the Carol Mann Agency for envisioning this project and making it happen.

Likewise, Carol Mann, there are not enough words to say how grateful I am for your belief in this project from the start and your willingness to take over as my agent when Lydia left the business. I could not ask for a bigger supporter in my corner. Thank you.

From the first time we met in his office, Stephen Morrow has been upbeat, enthusiastic, and excited about this project, sometimes even more than I could believe. His encouragement and support were unending throughout this entire process. But even more than that, he had a clear vision for what he thought this book could be, and I always felt like I was in great hands with Stephen as an editor. I am so thankful for his wit, creativity, devotion, and friendship.

Additionally, I cannot say thank you enough to the amazing team at Dutton whose hard work and dedication enabled this book to come to fruition. Madeline Newquist, thank you for answering all my first-time-author questions. Daniel Lagin, thank you for making complex scientific aspects come to life with your beautiful illustrations. Jennifer Eck, thank you for your exquisite copy editing, and Emily Canders, for your tireless marketing and publicity efforts.

I had an extraordinary fact-checker, Matt Mahoney, who continually exceeded my expectations with his curiosity, diligence, and support. You made this book better. Thank you.

I have said often that my job in this was easy. I just ask the questions. The scientists are the ones conducting the research, drafting the proposals, putting their reputations and careers on the line, and then, on top of it, they have to patiently answer my fumbling queries. I spent many hours listening to and learning from the incredible researchers that appear in this book, too many of them to individually list here. I hope they know how appreciative I am of their willingness to help in this endeavor, and I hope that you, the reader, could recognize their effort as you have gone through these pages.

I do have to extend a special shout-out to Jason Sherwin and Jordan Muraskin, a couple of true mensches, for allowing me into their lives, both personal and professional, for much longer than I probably deserved. Their openness and willingness to let me cover their progress, throughout its ups and downs, were the kinds of gifts that journalists dream about. Thanks for the laughs and the mystery meat.

I am endlessly grateful to Jason Stallman, sports editor at *The New York Times*, for allowing me to do what I love for so many years. Likewise, over the last six years I have been incredibly fortunate to work with so many great editors, including Jay Schreiber, Connor Ennis, Naila-Jean Meyers, Andrew Das, Melissa Hoppert, Bob Goetz, Carl Nelson, and Randy Archibold. Every day you make me a better writer and reporter. Terri Ann Glynn, thank you for making everything possible.

From my earliest reporting days at *The Daily Orange* to Columbia Journalism School, I have been fortunate to be surrounded by great teachers and mentors. Thank you to Pete Thamel and Greg Bishop, who took me under their wing and showed me how to do this job at the highest level. I still learn from reading Ben Shpigel and Zach Berman every day. Thank you, Julie Reiff, for giving me the journalism bug at Taft and remaining such a supportive voice. To Kevin Coyne, Dave Blum, and Sandy Padwe: your advice over the years and particularly during this project has been invaluable. Thank you to Roya Hakakian for her guidance and wisdom demystifying the book industry. Karol Kamin, thank you for being a fount of great advice and for being an early believer in this book (and me).

Thanks as well to Glenn Stout, who set this whole thing in motion. You gave me the opportunity to start this journey with Jason and Jordan not knowing where it could lead. For that, I am truly grateful.

Writing a neuroscience book, you spend a lot of time in libraries. Fortunately, I had access to some of the best. Thank you to the helpful staff at Columbia University's Butler Library, who fetched many interlibrary loan requests, as well as Rupert Baker and the research staff at the archives of the Royal Society and the Wellcome Library, in London. Mary Buckett in Cleveland and Anne Holden at Columbia, thank you for responding to every inquiry with a smile and an uncommon degree of helpfulness, which was so very appreciated.

I cannot forget the one and only Bill Kochevar, who graciously allowed me to hang out with him for a day and ask questions I'm sure he has answered a thousand times. Thank you and go Browns!

I am incredibly lucky to have such supportive and dedicated friends. Scott Cacciola, Matt Gelb, Tim Bella, thank you for always being there when I needed a read (or a beer). Yuval Brokman, thank you for being my personal photographer. Additional thanks to Sarah Brokman, for the cake.

Lastly, thank you to my entire family for championing me and this project from the outset. My parents, Julie and Roy; my sister, Steph, and soon-to-be brother-in-law, Matt; my in-laws Moish, Audrey, brother-in-law Lee, and sister-in-law Jess; and especially my grandmother, Nanny Jackie—your unwavering support, encouragement, and love made this possible. And, of course, to my wife, Missy, for reading, listening, brainstorming, reaffirming, and everything else. I love you.

ON SOURCES

For much of the information recorded in this book, I relied heavily on two sources: *The Principles of Psychology* by William James and *Principles of Neural Science*, 5th edition, edited by Eric R. Kandel, James H. Schwartz, Thomas M. Jessell, Steven A. Siegelbaum, and A. J. Hudspeth. The latter is a textbook the size of a toaster oven. It sat on the corner of my desk for a year. To understand the brain, its structure and function, you need not have to go much further than the pages contained within. But, of course, I did try to go further. I talked to the men and women currently on the front lines of motor research. A lot of information was gleaned from those conversations, beyond the comments I have quoted. I mention this knowing that neuroscience research can be both imperfect and impermanent, and few theories are ever agreed upon by everyone. I chose to focus on researchers who seemed to be challenging certain conventions or approaching difficult questions in novel ways. I attempted to strike a balance between what was interesting, what was relevant and what was significant, while recognizing that I could not possibly hit on every topic or shine a light down every avenue. There are 86 billion neurons in the brain, one for every academic paper written about it, or so it seemed. In my judgment, this book is meant to give a peek at the scope of the research into motor skills from the past to the present. It should not be confused for a textbook or an encyclopedia, of which there are plenty already in circulation. For those interested in tracing my footsteps, I have provided a list of relevant materials for every chapter. I must warn you, though, once you begin, it is hard to inhibit your neurons from firing.

SELECTED BIBLIOGRAPHY

1. DECERVO

Abernethy, B., & Russell, D. G. (1987). Expert–novice differences in an applied selective attention task. *Journal of Sport & Exercise Psychology, 9*(4), 326–345. https://doi.org/10.1123/jsp.9.4.326

Bartlett, F. C., & Burt, C. (1933). Remembering: A study in experimental and social psychology. *British Journal of Educational Psychology, 3*(2), 187–192.

Blakeslee, S., & Blakeslee, M. (2008). *The body has a mind of its own: How body maps in your brain help you do (almost) everything better.* New York, NY: Random House.

Claxton, G. (2015). *Intelligence in the flesh: Why your mind needs your body much more than it thinks.* New Haven, CT: Yale University Press.

James, W. (1950). *The principles of psychology* (Vol. 1). New York, NY: Dover Publications.

Kandel, E. R., Schwartz, J. H., Jessell, T. M., Siegelbaum, S. A., & Hudspeth, A. J. (Eds.). (2013). *Principles of neural science.* New York, NY: McGraw-Hill Medical.

Klawans, H. L. (1996). *Why Michael couldn't hit: And other tales of the neurology of sports.* New York, NY: Macmillan.

Lazenby, R. (2014). *Michael Jordan: The life.* New York, NY: Little, Brown.

Milton, J., Solodkin, A., & Small, S. L. (2008). Why did Casey strike out? The neuroscience of hitting. In D. Gordon (Ed.), *Your brain on Cubs: Inside the heads of players and fans.* New York, NY: Dana Press.

Muraskin, J. (2015). *Using neuroimaging to investigate the effect of expertise in rapid perceptual decision making* (doctoral dissertation). Columbia University, New York, NY.

Muraskin, J., Dodhia, S., Lieberman, G., Garcia, J. O., Verstynen, T., Vettel, J. M., . . . Sajda, P. (2016). Brain dynamics of post-task resting state are influenced by expertise: Insights from baseball players. *Human Brain Mapping, 37*(12), 4454–4471. doi:10.1002/hbm.23321

Muraskin, J., Sherwin, J., & Sajda, P. (2013). *A system for measuring the neural correlates of baseball pitch recognition and its potential use in scouting and player development.* Presented at MIT Sloan Sports Analytics Conference, Boston, MA.

Muraskin, J., Sherwin, J., & Sajda, P. (2015). Knowing when not to swing: EEG evidence that enhanced perception–action coupling underlies baseball batter expertise. *NeuroImage, 123*, 1–10. doi:10.1016/j.neuroimage.2015.08.028

Nakata, H., Yoshie, M., Miura, A., & Kudo, K. (2010). Characteristics of the athletes' brain: Evidence from neurophysiology and neuroimaging. *Brain Research Reviews, 62*(2), 197–211. doi:10.1016/j.brainresrev.2009.11.006

Robbins, J. (2000). *A symphony in the brain: The evolution of the new brain wave biofeedback.* New York, NY: Atlantic Monthly Press.

Sherwin, J., Muraskin, J., & Sajda, P. (2012). You can't think and hit at the same time: Neural correlates of baseball pitch classification. *Frontiers in Neuroscience, 6*, 177. doi:10.3389/fnins.2012.00177

Watts, R. G., & Bahill, T. (1990). *Keep your eye on the ball: The science and folklore of baseball.* New York, NY: W. H. Freeman.

2. THE MOVEMENT CHAUVINIST

Bernstein, N. A. (1967). *The co-ordination and regulation of movements.* Oxford, UK: Pergamon Press.

Cohen, R. G., & Sternad, D. (2012). State space analysis of timing: Exploiting task redundancy to reduce sensitivity to timing. *Journal of Neurophysiology, 107*(2), 618–627. doi:10.1152/jn.00568.2011

Epstein, D. (2014). *The sports gene: Inside the science of extraordinary athletic performance.* New York, NY: Penguin.

Faisal, A. A., Selen, L. P. J., & Wolpert, D. M. (2008). Noise in the nervous system. *Nature Reviews Neuroscience, 9*(4), 292–303. doi:10.1038/nrn2258

Franklin, D. W., & Wolpert, D. M. (2011). Computational mechanisms of sensorimotor control. *Neuron 72*(3), 425–442. doi:10.1016/j.neuron.2011.10.006

Harris, C. M., & Wolpert, D. M. (1998). Signal-dependent noise determines motor planning. *Nature, 394*(6695), 780–784. doi:10.1038/29528

Huber, M. E., Kuznetsov, N., & Sternad, D. (2016). Persistence of reduced neuromotor noise in long-term motor skill learning. *Journal of Neurophysiology, 116*(6), 2922–2935. doi:10.1152/jn.00263.2016

Jones, K. E., Hamilton, A. F. C., & Wolpert, D. M. (2002). Sources of signal-dependent noise during isometric force production. *Journal of Neurophysiology, 88*(3), 1533–1544.

Kording, K. P., & Wolpert, D. M. (2004). Bayesian integration in sensorimotor learning. *Nature, 427*(6971), 244–247. doi:10.1038/nature02169

Latash, M. L., & Zatsiorsky, V. M. (2001). *Classics in movement science.* Champaign, IL: Human Kinetics.

Lewis, M. (2016). *The undoing project: A friendship that changed the world.* London: Penguin UK.

Mook, D. G. (2004). *Classic experiments in psychology.* Westport, CT: Greenwood Publishing.

Müller, H., & Sternad, D. (2004). Decomposition of variability in the execution of goal-oriented tasks: Three components of skill improvement. *Journal of Experimental Psychology: Human Perception and Performance 30*(1), 212. doi:10.1037/0096-1523.30.1.212

Sternad, D., Abe, M. O., Hu, X., & Müller, H. (2011). Neuromotor noise, error tolerance and velocity-dependent costs in skilled performance. *PLoS Computational Biology, 7*(9). doi:10.1371/journal.pcbi.1002159

Wilkinson, M. (2016). *Restless creatures: The story of life in ten movements.* London, UK: Icon Books.

Wolpert, D. M., & Flanagan, J. R. (2001). Motor prediction. *Current Biology, 11*(18), R729–R732. doi:10.1016/S0960-9822(01)00432-8

3. THE MOTOR HUNTER

Dudman, J. T., & Krakauer, J. W. (2016). The basal ganglia: From motor commands to the control of vigor. *Current Opinion in Neurobiology, 37,* 158–166. doi:10.1016/j.conb.2016.02.005

Ericsson, K. A. (2017). *Peak: Secrets from the new science of expertise.* Boston, MA: Mariner Books.

Herculano-Houzel, S. (2017). *The human advantage: How our brains became remarkable.* Cambridge, MA: MIT Press.

Mazzoni, P., Hristova, A., & Krakauer, J. W. (2007). Why don't we move faster? Parkinson's disease, movement vigor, and implicit motivation. *The Journal of Neuroscience, 27*(27), 7105–7116. doi:10.1523/jneurosci.0264-07.2007

Milner, B., Corkin, S., & Teuber, H. L. (1968). Further analysis of the hippocampal amnesic syndrome: 14-year follow-up study of HM. *Neuropsychologia, 6*(3), 215–234. doi:10.1016/0028-3932(68)90021-3

Sagan, C. (1977). *The dragons of Eden: Speculations on the evolution of human intelligence.* New York, NY: Ballantine.

Shadmehr, R., & Krakauer, J. W. (2008). A computational neuroanatomy for motor control. *Experimental Brain Research, 185*(3), 359–381. doi:10.1007/s00221-008-1280-5

Stanley, J., & Krakauer, J. W. (2013). Motor skill depends on knowledge of facts. *Frontiers in Human Neuroscience, 7.* doi:10.3389/fnhum.2013.00503

Taylor, J. A., Krakauer, J. W., & Ivry, R. B. (2014). Explicit and implicit contributions to learning in a sensorimotor adaptation task. *The Journal of Neuroscience, 34*(8), 3023–3032. doi:10.1523/jneurosci.3619-13.2014

Williams, A. M., Davids, K., & Williams, J. G. P. (1999). *Visual perception and action in sport*. London, UK: Taylor & Francis, pp. 124–143.

Wong, A. L., & Haith, A. M. (2017). Motor planning flexibly optimizes performance under uncertainty about task goals. *Nature Communications, 8*, 14624. doi:10.1038/ncomms14624

Wong, A. L., Haith, A. M., & Krakauer, J. W. (2015). Motor planning. *The Neuroscientist, 21*(4), 385–398. doi:10.1177/1073858414541484

Yarrow, K., Brown, P., & Krakauer, J. W. (2009). Inside the brain of an elite athlete: The neural processes that support high achievement in sports. *Nature Reviews Neuroscience, 10*(8), 585–596. doi:10.1038/nrn2672

4. "FROM MIND TO MUSCLE"

Bailey, C. W., & Darwin, B. (1924). *The brain and golf: Some hints from modern mental science*. Boston, MA: Small.

Eccles, J., & Gibson, W. C. (1979). *Sherrington: His life and thought*. Berlin, Germany: Springer-Verlag.

Ferrier, D. (1874). On the localisation of the functions of the brain. *British Medical Journal, 2*(729), 766.

Finger, S. (2000). *Minds behind the brain: A history of the pioneers and their discoveries*. New York, NY: Oxford University Press.

Foster, M. (1897). *A Text-Book of Physiology, Part III*. London, UK: Macmillan, pp. 922–1001.

Foster, M. (1901). *Lectures on the history of physiology during the sixteenth, seventeenth, and eighteenth centuries*. Cambridge, UK: Cambridge University Press.

Fritsch, G., & Hitzig, E. (2009). Electric excitability of the cerebrum (*Über die elektrische Erregbarkeit des Grosshirns*). *Epilepsy & Behavior, 15*(2), 123–130. doi:10.1016/j.yebeh.2009.03.001

Gross, C. G. (1995). Aristotle on the brain. *The Neuroscientist, 1*(4), 245–250.

Swazey, J. (1969). *Reflexes and motor integration: Sherrington's concept of motor integration*. Cambridge, MA: Harvard University Press.

Sherrington, C. E. R., & Sherrington, C. S. (1957). *Charles Scott Sherrington, 1857–1952: Memories*. London, UK: The Royal Society.

Sherrington, C. S. (1946). *The endeavor of Jean Fernel, with a list of his writings*. Cambridge, UK: Cambridge University Press.

Sherrington, C. S. (2009). *Man on his nature*. Cambridge, UK: Cambridge University Press.

Taylor, C. S. R., & Gross, C. G. (2003). Twitches versus movements: A story of motor cortex. *The Neuroscientist, 9*(5), 332–342.

Zimmer, C. (2005). *Soul made flesh: The discovery of the brain—and how it changed the world*. New York, NY: Simon & Schuster.

SELECTED BIBLIOGRAPHY

5. THE NEUROTECH SPACE

MacKinnon, J. B. (2016, August 11). The strange brain of the world's greatest solo climber. *Nautilus, 19.* http://nautil.us/issue/39/sport/the-strange-brain-of-the-worlds-greatest-solo-climber

Naito, E., & Hirose, S. (2014). Efficient foot motor control by Neymar's brain. *Frontiers in Human Neuroscience, 8.* doi:10.3389/fnhum.2014.00594

Schonbrun, Z. (2017, January 8). Honing the "gymnastics of the brain." *The New York Times* (New York ed.), p. SP1.

6. SEARCHING FOR THE MOTOR ENGRAM

Adams, J. A. (1987). Historical review and appraisal of research on the learning, retention, and transfer of human motor skills. *Psychological Bulletin, 101*(1), 41.

Adrian, E. D. (1943). Sensory areas of the brain. *The Lancet, 242*(6254), 33–35.

Ajemian, R., D'Ausilio, A., Moorman, H., & Bizzi, E. (2010). Why professional athletes need a prolonged period of warm-up and other peculiarities of human motor learning. *Journal of Motor Behavior, 42*(6), 381–388. doi:10.1080/00222895.2010.528262

Beach, F. A. (1961). *Karl Spencer Lashley.* Washington, DC: National Academy of Sciences.

Bernstein, N. (1935). The problem of the interrelation of coordination and localization. *Archives of Biological Sciences, 38,* 15–59.

Calvin, W. H. (1983). *The throwing Madonna: Essays on the brain.* New York, NY: McGraw-Hill.

Grafton, S. T., & Hamilton, A. F. C. (2007). Evidence for a distributed hierarchy of action representation in the brain. *Human Movement Science, 26*(4), 590–616. doi:10.1016/j.humov.2007.05.009

Guo, J.-Z., Graves, A. R., Guo, W. W., Zheng, J., Lee, A., Rodríguez-González, J., . . . Hantman, A. W. (2015). Cortex commands the performance of skilled movement. *eLife, 4,* e10774. doi:10.7554/elife.10774

Lashley, K. (1951). The problem of serial order in behavior. In L. A. Jeffress (Ed.), *Cerebral mechanisms in behavior.* Oxford, UK: Wiley, pp. 112–136.

Lieberman, D. (2014). *The story of the human body: Evolution, health, and disease.* New York, NY: Vintage.

Linden, D. J. (2016). *Touch: The science of hand, heart, and mind.* New York, NY: Penguin.

Penfield, W., & Rasmussen, T. (1950). *The cerebral cortex of man: A clinical study of localization of function.* Oxford, UK: Macmillan.

Pruszynski, J. A., & Johansson, R. S. (2014). Edge-orientation processing in first-order tactile neurons. *Nature Neuroscience, 17*(10), 1404–1409. doi:10.1038/nn.3804

Schmidt, R. A. (1975). A schema theory of discrete motor skill learning. *Psychological Review, 82*(4), 225.

Watson, A. H. D. (2006). What can studying musicians tell us about motor control of the hand? *Journal of Anatomy, 208*(4), 527–542. doi:10.1111/j.1469-7580.2006.00545.x

Wilson, F. R. (2010). *The hand: How its use shapes the brain, language, and human culture.* New York, NY: Vintage.

7. EMBODIED EXPERTISE

Aglioti, S. M., Cesari, P., Romani, M., & Urgesi, C. (2008). Action anticipation and motor resonance in elite basketball players. *Nature Neuroscience, 11*(9), 1109–1116. doi:10.1038/nn.2182

Bläsing, B., Calvo-Merino, B., Cross, E. S., Jola, C., Honisch, J., & Stevens, C. J. (2012). Neurocognitive control in dance perception and performance. *Acta Psychologica, 139*(2), 300–308. doi:10.1016/j.actpsy.2011.12.005

Brown, S., & Parsons, L. M. (2008). The neuroscience of dance. *Scientific American, 299*(1), 78–83.

Calvo-Merino, B., Glaser, D. E., Grèzes, J., Passingham, R. E., & Haggard, P. (2004). Action observation and acquired motor skills: An FMRI study with expert dancers. *Cerebral Cortex 15*(8), 1243–1249. doi:10.1093/cercor/bhi007

Clarey, C. (2014, February 23). Olympians use imagery as mental training. *The New York Times* (New York ed.), p. SP1.

Cross, E. S., Hamilton, A. F. C., & Grafton, S. T. (2006). Building a motor simulation de novo: Observation of dance by dancers. *Neuroimage, 31*(3), 1257–1267. doi:10.1016/j.neuroimage.2006.01.033

Di Pellegrino, G., Fadiga. L., Fogassi, L., Gallese, V., & Rizzolatti, G. (1992). Understanding motor events: A neurophysiological study. *Experimental Brain Research, 91*(1), 176–180.

Eaves, D. L., Riach, M., Holmes, P. S., & Wright, D. J. (2016). Motor imagery during action observation: A brief review of evidence, theory and future research opportunities. *Frontiers in Neuroscience, 10.* doi:10.3389/fnins.2016.00514

Kirsch, L. P., & Cross, E. S. (2015). Additive routes to action learning: Layering experience shapes engagement of the action observation network. *Cerebral Cortex, 25*(12), 4799–4811. doi:10.1093/cercor/bhv167

Kirsch, L. P., Dawson, K., & Cross, E. S. (2015). Dance experience sculpts aesthetic perception and related brain circuits. *Annals of the New York Academy of Sciences, 1337*(1), 130–139. doi:10.1111/nyas.12634

Kirsch, L. P., Urgesi, C., & Cross, E. S. (2016). Shaping and reshaping the aesthetic brain: Emerging perspectives on the neurobiology of embodied aesthetics. *Neuroscience & Biobehavioral Reviews, 62,* 56–68. doi:10.1016/j.neubiorev.2015.12.005

Spackman, K. (2009). *The winner's bible: Rewire your brain for permanent change.* Austin, TX: Greenleaf Book Group.

Sperry, R. W. (1952). Neurology and the mind-brain problem. *American Scientist, 40*(2), 291–312.

Uithol, S., Van Rooij, I., Bekkering, H., & Haselager, P. (2011). Understanding motor resonance. *Social Neuroscience, 6*(4), 388–397. doi:10.1080/17470919.2011.559129

8. THE BODY IN SPACE

Blakemore, S.-J., Wolpert, D., & Frith, C. (2000). Why can't you tickle yourself? *Neuroreport, 11*(11), R11–R16.

Burdick, A. (2017). *Why time flies: A mostly scientific investigation.* New York, NY: Simon & Schuster.

Dash, S., Yan, X., Wang, H., & Crawford, J. D. (2015). Continuous updating of visuospatial memory in superior colliculus during slow eye movements. *Current Biology, 25*(3), 267–274. doi:10.1016/j.cub.2014.11.064

Dennett, D. (1991). *Consciousness: Explained.* Boston, MA: Little, Brown.

Eagleman, D. M. (2009). Brain time. In M. Brockman (Ed.), *What's next: Dispatches from the future of science.* New York, NY: Vintage, pp. 155–169.

Eagleman, D. M., & Sejnowski, T. J. (2000). Motion integration and postdiction in visual awareness. *Science, 287*(5460), 2036–2038.

Han, J., Waddington, G., Anson, J., & Adams, R. (2015). Level of competitive success achieved by elite athletes and multi-joint proprioceptive ability. *Journal of Science and Medicine in Sport, 18*(1), 77–81. doi:10.1016/j.jsams.2013.11.013

Klier, E. M., & Angelaki, D. E. (2008). Spatial updating and the maintenance of visual constancy. *Neuroscience 156*(4), 801–818. doi: 10.1016/j.neuroscience .2008.07.079

Medendorp, W. P., Goltz, H. C., Vilis, T., & Crawford, J. D. (2003). Gaze-centered updating of visual space in human parietal cortex. *Journal of Neuroscience, 23*(15), 6209–6214.

Miall, R. C., & Wolpert, D. M. (1996). Forward models for physiological motor control. *Neural Networks, 9*(8), 1265–1279.

Moser, E. I., Kropff, E., & Moser, M. B. (2008). Place cells, grid cells, and the brain's spatial representation system. *Annual Review of Neuroscience, 31*, 69–89. doi: 10.1146/annurev.neuro.31.061307.090723

O'Keefe, J., & Nadel, L. (1978). *The hippocampus as a cognitive map.* Oxford, UK: Clarendon Press.

Sacks, O. (2009). *The man who mistook his wife for a hat: And other clinical tales.* New York, NY: Picador.

Shergill, S. S., Bays, P. M., Frith, C. D., & Wolpert, D. M. (2003). Two eyes for an eye: The neuroscience of force escalation. *Science, 301*(5630), 187. doi:10.1126 /science.1085327

Sherrington, C. S. (1909). On plastic tonus and proprioceptive reflexes. *Quarterly Journal of Experimental Physiology, 2*(2), 109–156.

Thorndike, E. L. (1901). The evolution of the human intellect. *Popular Science Monthly, 60.*

Tolman, E. C. (1948). Cognitive maps in rats and men. *Psychological Review, 55*(4), 189–208.

Wickens, A. P. (2015). *A history of the brain: From Stone Age surgery to modern neuroscience.* New York, NY: Psychology Press, p. 202.

Wolpert, D. M., Ghahramani, Z., & Jordan, M. I. (1995). An internal model for sensorimotor integration. *Science, 269*(5232), 1880–1882.

9. A PARALYZED MAN WHO MOVED

Ajiboye, A. B., Willett, F. R., Young, D. R., Memberg, W. D., Murphy, B. A., Miller, J. P., . . . Kirsch, R. F. (2017). Restoration of reaching and grasping movements through brain-controlled muscle stimulation in a person with tetraplegia: A proof-of-concept demonstration. *The Lancet. 389*(10081), 1821–1830. doi:10.1016/s0140-6736(17)30601-3

Evarts, E. V. (1968). A technique for recording activity of subcortical neurons in moving animals. *Electroencephalography and Clinical Neurophysiology, 24*(1), 83–86.

Georgopoulos, A. P., Caminiti, R., Kalaska, J. F., & Massey, J. T. (1983). Spatial coding of movement: A hypothesis concerning the coding of movement direction by motor cortical populations. *Experimental Brain Research, 49*(Suppl. 7), 327–336.

Squire, L. R. (Ed.). (1996). *The history of neuroscience in autobiography* (Vol. 1). Washington, DC: Society for Neuroscience, p. 340.

Thach, W. T. (2000). Edward Vaughan Evarts. In *Biographical Memoirs* (Vol. 78). Washington, DC: National Academies Press, p. 34.

INDEX

INDEX

INDEX

Laby, Daniel, 26–27
Lancet (journal), on BrainGate, 296
LaRitz, Tom, 21
Lashley, Karl, 190–192, 203–204
lateralization, 196–198, 213–214
Leeuwenhoek, Antoni van, 123–124
Lemon, Roger, 194
Leonard, Kawhi, 209, 209n
Leonardo Da Vinci, 45, 121
Libet, Benjamin, 274
"Limits to Human Performance"
    workshop, 77–82
Linden, David, 56–57, 218–219, 221
Locke, John, 69
Longet, François, 129
Lotze, Hermann, 232

MacDonald, John, 75
Maguire, Eleanor, 24–25
Major League Baseball. *See* baseball
    teams and batters
Man on His Nature (Sherrington), 151
Mantle, Mickey, 17
Matteucci, Carlo, 129
Mays, Willie, 17
Mazzoni, Pietro, 106, 107, 114
McGurk, Harry, 75
McGurk effect, 75, 75n
McNally, Kat, 100
McNamee, Dan, 63, 65
mechanoreceptors, 214–220, 222–223,
    263, 266, 267–268
Memberg, Bill, 300, 301, 302
Messi, Lionel, 62–63
Miles, Walter, 20
Miller, Lee, 308
Milner, Brenda, 84, 287
mirror neurons, 234–237, 239, 243–244
MIT Sloan Sports Analytics Conference,
    153–156, 180, 184–185
Molaison, Henry, 83–85, 91–92, 93, 94, 287
Mondino de Luzzi, 121
Moser, Edvard, 288
motivation, 65, 97–103, 113–114
motor cortex
    debate on (1881), 132–135
    electrical stimulation of, 126–127,
        131–132
    mapping of, 128–129, 132, 198–201

mind-body connection and,
    151–152, 200
movement initiation, 3, 33, 39–41,
    119–121, 177, 229, 254–255, 256–257
motor engram, 187–224
    chunking and, 201–214. *See also*
        chunking
    defined, 191, 202
    hand dexterity and, 193–196
    historical context for, 187–192,
        193–197
    lateralization and, 196–198, 213–214
    mechanoreceptors, 214–220, 222–223,
        263, 266, 267–268
    motor cortex mapping, 198–201
    sensorimotor control and, 218–224
motor resonance, 240–241, 246–253
motor system
    control of, 47–76, 119–152, 187–224.
        See also control of motor system;
        motor engram
    critique of current research, 48–49,
        80–82, 106–110
    overview, 3–5
    proprioception and, 259–291. *See also*
        proprioception
    skills and, 77–117. *See also* skills
Mountcastle, Vernon, 233–234, 271–272
movement, neuroscience of. *See*
    performance, neuroscience of
movement, robotic, 49–50, 105, 221
Muraskin, Jordan
    background, 30–34, 156–158
    on deCervo's viability, 175–176
    on error processing, 253–255
    Krakauer on, 109
    on marketing deCervo, 153–155,
        158–166, 171–175, 184–185, 312–313
    on swing decisions quantification,
        7–10, 15–17, 36–38, 39–45, 166–170,
        179–181
    vision for deCervo, 2–3, 14–16, 17,
        44–45, 185
Murphy, Brian, 297–299, 301
muscle memory, 11, 41, 89–91,
    189–190, 291
muscle spindles, 263, 266, 267–268
musicians, 33, 66, 211–213, 212n, 222,
    223–224, 255, 312

INDEX

Philosophical Investigations
(Wittgenstein), 95
*Philosophical Transactions of the Royal
Society of London* (journal), on
Bayesian theory, 70
Phoenix Suns, 222
phrenology, 127–128
physical therapy, 98–103, 251–252
pianists, 66, 211–212, 223–224
pitches. *See* baseball teams and batters
place cells, 287–288, 290–291
plasticity, 22–24, 101–102, 127, 208,
252, 305
Plato, 84, 120
play, as motivation, 99–103
Pomeranz, Drew, 181
Popp, Nicola, 209–210
Posit Science, 182n
Posner, Michael, 89
predictions
Bayesian theory of, 70–76, 70n, 216,
240–241, 252–253
motor experience used for, 229–231
sensorimotor, 41, 69, 112–113, 271–273,
276–283
during swing decisions, 21–29, 39,
41–42, 164–165, 229
prefrontal cortex, 38
Preller, A. J., 180–181
Pride and a Daily Marathon (Cole), 264
Principles of Neural Science (Kandel,
Schwartz, Jessell, eds.), 4–5,
104–105, 321
Principles of Psychology, The (James), 131,
264, 321
prior knowledge, 22, 71–73, 229–231,
238–239, 252
probability, conditional, 69–76, 70n, 216,
240–241, 252–253
procedural knowledge, 84–85, 87–88,
91–93
Prokopowicz, Bob, 29–30
proprioception, 259–291
activity studies on, 263–264
defined, 263
disorders of, 264–267
historical context, 259–263
neural communication of, 267–268
sensorimotor predictions, 278–283

sensory feedback delays, 216–217,
274–278, 282
spatial updating and, 268–273
visuospatial reckoning, 283–291
Pruszynski, Andrew, 215–216, 217–219,
223–224, 267–268
Purkinje, Jan Evangelista, 136
Pylyshyn, Zenon W., 181

race car drivers, 225–228, 229, 257–258
Rafferty, Mary, 131–132
Ramachandran, Vilayanur, 235
Ramirez, Manny, 26–27
rapid perceptual decision-making
of athletes, 2, 28–29, 43–45, 59,
229–231
optimization of, 12–16, 73, 115–116,
164–169. *See also* Bayesian theory
and equation; deCervo
timing of, 274–278, 313
Rasmussen, Theodore, 198–201
reaction times, 12–13, 20, 110–116,
226–227, 311–313
reciprocal innervation, 144–146
reflexes, 112, 139–140, 142–149, 150–151,
189–190, 265
Rizzolatti, Giacomo, 236, 252
robotic movement, 49–50, 105, 221
rock climbers, 178
Rodgers, Aaron, 220
Ronaldo, Cristiano, 62–63
roof-brain, 151–152. *See also* motor cortex
Roy, Promit, 100, 101
rugby players, 289–290
runners, 68, 68n, 78–79
Russell, D. G, 21
Ruth, Babe, 11, 26, 26n

saccades, 270–273
Sajda, Paul, 30, 32–33, 38–39, 40
San Antonio Spurs, 208–209
Sandlin, Destin, 86–88
scaffolding, 74, 87, 92
Schelling, Xavi, 177
Schiff, Moritz, 129
Schmidt, Richard, 201, 207
Schueler, Ron, 19
Science (magazine), on myelination, 56n
Science of Hitting (Williams), 21–22

338

Printed in the United States
by Baker & Taylor Publisher Services.

Printed in the United States
by Baker & Taylor Publisher Services